The Living World of Audubon Mammals

The Living World of AUD

The photographers of The Living World of Audubon Mammals include:

Erwin A. Bauer Bill Browning Tom Brakefield

Ed Cesar Ed Cooper Helen Cruickshank Thase Daniel Harry Engels

Kenneth Fink S. J. Krasemann John R. MacGregor

Anthony Mercieca Charles E. Mohr Charles Ott Leonard Lee Rue III

Leonard Lee Rue IV *Alvin E. Staffan Marty Stouffer*

Arthur Swoger Larry West Michael Wotton

designed by Albert Squillace

Mammals

by Robert Elman

A Ridge Press Book / Grosset & Dunlap, Publishers, New York
A FILMWAYS COMPANY

Editor-in-chief: Jerry Mason
Editor: Adolph Suehsdorf
Art Director: Albert Squillace
Associate Editor: Ronne Peltzman
Associate Editor: Joan Fisher
Art Associate: Nancy Louie
Art Associate: David Namias
Art Production: Doris Mullane

The Living World of Audubon Mammals, by Robert Elman

Prepared and produced by The Ridge Press, Inc.
Published in 1976 by Grosset & Dunlap, Inc.
Published simultaneously in Canada.
Library of Congress Catalog Card Number: 76-432
ISBN: 0-448-12459-9
Printed and bound in Italy by Mondadori Editore, Verona

To my mother

Contents

Introduction

"The animals have never been carefully described," wrote the Reverend John Bachman in a letter to his friend John James Audubon, "and you will find difficulties at every step. Books cannot aid you much. Long journeys will have to be undertaken."

A celebrated amateur naturalist, the Reverend Mr. Bachman probably knew as much about American mammals as Audubon knew about birds, but whereas Audubon was a restless, impulsive man of action, Bachman was a cautious, methodical investigator. He had no wish to dissuade the great ornithologist from attempting a proposed work on quadrupeds which, in accuracy, comprehensiveness, and artistry, would stand with the newly completed prints of *The Birds of America* and the accompanying five volumes of *Ornithological Biographies.* What he wanted was to help assure success in this herculean exploration of the unknown by offering counsel. Citing examples of the obstacles ahead, he warned that the western deer would be troublesome to classify and the "ever varying Squirrels seem sent by Satan himself, to puzzle the Naturalists . . . . Say in what manner I can assist you."

Audubon's reply was to invite Bachman to share the ambitious task, thus initiating one of the more remarkable and important collaborations in the history of American field studies. As for those satanic members of the order *Rodentia,* the clergyman wrestled with them successfully, ultimately recognizing and including two dozen varieties of tree squirrels in Audubon's huge and unprecedented work, *The Viviparous Quadrupeds of North America.*

Perhaps only a wilderness adventurer of Audubon's experience, obstinacy, and self-regard would have been eager to undertake the necessary explorations at that time, when much of the West was unmapped and an expedition might be brought to a calamitous end by thirst or flood, heat or cold, storms, impassable terrain, lack of forage, or hostile Indians. His first tongue was French, his upbringing that of a wealthy, pampered colonial sent home to acquire the refinements of an aristocratic civilization; but by nature he was, as he styled himself, an "American woodsman."

His father was Jean Audubon, a naval lieutenant who had become a wealthy sugar planter and slave trader in the French West Indies. The boy John James, nee Fougère, was one of an undetermined number of bastard children the lieutenant sired with two Creole mistresses. He was born at Les Cayes on Saint-Domingue—now Haiti—on April 26, 1785. Within a year his mother died, and in 1789 his father took him and his infant half-sister Muguet to France, where Lieutenant Audubon's childless wife patiently awaited. Five years later Lieutenant and Madame Audubon formally adopted both children, whose paternity had been legally acknowledged. The boy was now called Jean Jacques Fougère Audubon, a name he later Americanized.

His tutors (and his father), though distressed by his mathematical ineptitude, were pleased by his aptitude for fencing, dancing, and playing the violin; and he was outrageously spoiled by his stepmother. Reminiscing years later, he recalled that she "hid my faults, boasted to everyone of my youthful merits, and said frequently in my presence that I was the handsomest boy in France."

In his adolescence he became an amateur taxidermist and an avid collector of bird nests and eggs. He also spent a great deal of time sketching wild creatures. When his father sent him to a naval school, hoping to interest him in something more "useful," the boy took leave through an open window and returned to the Audubon estate near Nantes. The elder Audubon, in resignation or despair, next sent his son to Paris to study at the art school of the famous painter Jacques Louis David. But the young wildlife portraitist felt stifled by rigidly academic art instruction. His Parisian sojourn was brief.

The plantation was lost by then, the West Indies beset by piracy and warfare, and Jean Audubon was no longer a very wealthy man, but he had purchased a lead mine and farm called Mill Grove in eastern Pennsylvania. To this farm in the New World he sent his ne'er-do-well son in 1803. The young man, now eighteen, felt an "indescribable pleasure" at the prospect of becoming "master of Mill Grove"—his description of himself despite the presence of a Quaker overseer named Miers Fisher, who was to supervise the Audubon scion as well as the Audubon holdings.

Soon John James Audubon was riding about the Pennsylvania countryside in new hunting clothes: black satin breeches "with silk stockings, and the finest ruffled shirt

Philadelphia could afford." He clashed with Miers Fisher over his neglect of the affairs of the mine and farm, but his days afield were not idle. Nature, particularly birds, had awakened an obsessive, lifelong curiosity in him, and evidently he already felt some inkling of the peculiar path his genius would take. Collecting specimens and meticulously drawing details of their anatomy and plumage, he began to pose them in lifelike attitudes, an original idea that had eluded most contemporary painters of wildlife.

While studying some pewees nesting in a cave, he found that by patience, keen observation, and intuitive understanding, he could overcome their timidity, virtually taming them. His perception of seasonal migrations was no less dim than that of other naturalists at the beginning of the nineteenth century, but he had the sense or curiosity to wonder if the birds would return to the same nesting area the next year. And he had the inspiration to find out by means of a simple experiment. As he afterward recalled, he "fixed a light silver thread to the leg of each." Thus, in April of 1804, John James Audubon performed the first bird-banding in America, probably the first successful experiments of this kind in the world. No other naturalist was sufficiently clever or creative to conduct such experiments until ninety-five years later, when a Danish schoolmaster named H. C. C. Mortensen, evidently unaware of Audubon's discoveries, banded teal, storks, and starlings.

At about the time of the pewee experiment Audubon met an aristocratic English-born sportsman named William Bakewell, who owned a large farm near Mill Grove, and fell in love with Bakewell's fifteen-year-old daughter Lucy. By then Miers Fisher had been replaced by one Francis Dacosta, who surpassed even Fisher in his disapproval of young Audubon's frivolities. Audubon, having proposed marriage to Lucy Bakewell, wished to visit France in order to obtain parental consent and financial aid. Dacosta refused him money for the voyage, but he was able to borrow it from a Bakewell relative. The voyage was a success. Consent and funds were forthcoming, as was an arrangement for a partnership with a young merchant, Ferdinand Rozier, who accompanied Audubon back to the United States.

Arriving in New York in the spring of 1806, Audubon and Rozier followed their fathers' instructions to gain commercial experience by working as clerks in a business establishment. However, Audubon frequently absented himself to continue collecting and drawing, and to study taxidermy with Samuel Mitchill, a naturalist. The brief New York mercantile period was portentous, for Audubon's only real business venture was a disastrous speculation in the failing indigo market. Shortly thereafter he and Rozier journeyed to Mill Grove, and that fall, at the District Court in Philadelphia, John James Audubon applied for and received American citizenship. He then sold the lead mine to Dacosta, rented his farmlands, and in 1807 set off with Rozier for Kentucky to open a general store in the frontier town of Louisville. The following year he returned, married Lucy Bakewell, and brought her back to Louisville. She was a remarkably compatible spouse for an impecunious, roving artist-naturalist. Her faith in his genius sustained her during his long absences while he traveled beyond the frontiers to study the country's wildlife, and she labored to support the family when he could not.

Even after Lucy bore his first son, Victor Gifford, in 1809, Audubon could not force himself to spend his days "in serious storekeeping" when the sounds of the forest and waters drifted through the windows. The next year, with business dwindling, the partners loaded their wares aboard a flatboat and floated a hundred and twenty-five miles down the Ohio to Henderson, Kentucky, where they opened another establishment. While Rozier toiled diligently in the store, Audubon took their only helper, a young clerk named John Pope, into the woods to collect wildlife specimens. The truants "roamed the country," according to one biographer, "in eager pursuit of rare birds, and with rod and gun bountifully supplied the table." Unfortunately, Audubon was able to supply little else for his household.

During this period, while traveling near Frankfort, Audubon met Daniel Boone. Then in his seventies but still hale and active, Boone conformed to Audubon's ideal of the woodsman, and Audubon was delighted when Boone took him squirrel hunting. Near the banks of the Kentucky River on land "thickly covered with black walnuts, oaks and hickories," Audubon wrote of the occasion, " . . . squirrels were seen gambolling on every tree around us." Although he was then concentrating his artistic and recording efforts on birds, he used some of the squirrels and other small game animals taken in his wanderings as models for sketching before using them again over his campfire. In *The Birds of America* his delicately kinesthetic barred owl—rendered with a poise and flow reminiscent of Oriental art—alights on a branch to strike at an apparently unconcerned eastern gray squirrel that is drawn as meticulously as the bird. The very same squirrel, traced and transposed, appears again in one of the prints in his mammalian work. "How often," he wrote, commenting on his encounters with owls in the *Biographies,* "when snugly settled under the boughs of my temporary encampment and preparing to roast a venison steak or the body of a squirrel, have I been saluted with the exulting bursts of this nightly disturber of the peace."

Acquaintance with men like Boone intensified Audubon's yearning to see more of America's wilderness and wildlife. When Rozier suggested moving on again, this time to the French settlement of Sainte Geneviève on the Mississippi, Audubon readily went with him, leaving his wife and child at a friend's farm near Henderson. The partners loaded a keelboat with three hundred barrels of whiskey, a commodity that usually brought a good profit in frontier towns even when other merchandise could hardly be bartered. Delayed by ice on the river, they stayed for a while at Indian encampments. Audubon sketched chalk portraits of Shawnees and Osages in exchange for wildlife specimens, and he continued to draw the birds and animals of the forests. Like George Catlin, another artist who was soon to travel westward, he considered the aborigines worthy of serious study, afterward describing in detail their manner of living and some of their customs. Evidently he regretted having to push on to Sainte Geneviève, although he achieved temporary solvency there when the three hundred barrels of whiskey, purchased for twenty-five cents a gallon, were quickly sold for eight times that much.

Rozier stayed at the settlement to open a new store, but

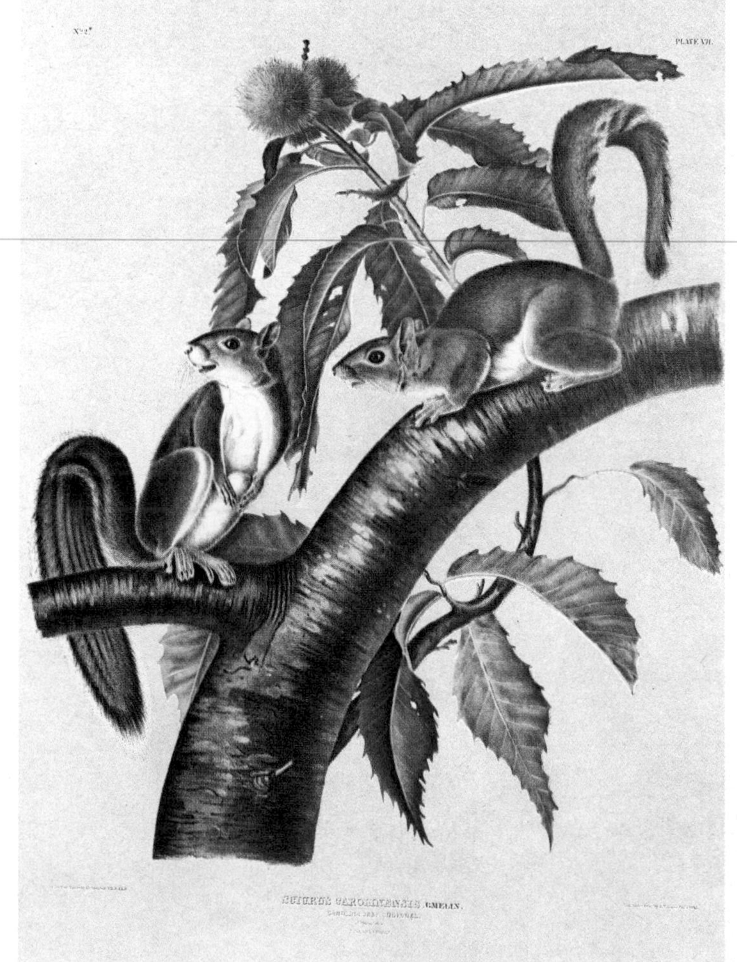

The eastern gray squirrel depicted with a barred owl in *The Birds of America* reappeared in the *Quadrupeds.* The earlier work classified the rodent as *sciurus cinereus,* literally meaning gray squirrel; the second called it *sciurus carolinensis,* the designation that is now accepted.

Audubon sold his share in the enterprise to his partner and took the overland trail back to Henderson, equipped only with a gun and a knapsack for his sketching materials and scanty provisions. Alone but for his dog, he sometimes camped in the woods, sometimes spent the night with Indians or in settlers' cabins.

By now he had made and saved over two hundred life-size drawings of wild creatures, chiefly birds but including a few small mammals and reptiles. They were not yet the finished portraits, done in a combination of watercolors, chalk, and crayon, from which lithographs were ultimately made for *The Birds of America* and *The Viviparous Quadrupeds of North America,* but they were the nucleus of the first work and the inspiration for the second. That he did not yet envisage the publication of his art is hard to understand, for by now, too, he realized that there was a great popular curiosity about wild creatures, that a man might even earn a livelihood by publishing pictures and descriptions of fauna and flora.

Private and public museums were being established. Specimen collecting was in fashion. Thomas Jefferson, himself an ardent collector, amateur natural scientist, and president of the American Philosophical Society, had instructed Lewis and Clark to gather natural specimens on their famous expedition into the territory of the Louisiana Purchase. Mark Catesby's *Natural History of Carolina,* a pioneering work of scientific illustration, was still famous some eighty years after publication, and other botanical and biological treatises were also in demand. John Bartram, America's first great botanist and, with his son William, a true founder of the science of natural interrelationships that would much later be named ecology, had built the nation's earliest botanical garden on the banks of the Schuylkill, near Philadelphia. Since John Bartram's death in 1777, his sons had continued to maintain the garden and advance the study of nature. In 1791 William Bartram had published his book, *Travels,* in which he described wild plants and animals of the South and the western frontier. He was celebrated as well for his records of bird sightings and migrations; indeed, his listing of two hundred and fifteen native birds was the most comprehensive until the publication of Alexander Wilson's monumental works.

Finally, there was Wilson himself, the eminent Scottish-born ornithologist who had been encouraged by William Bartram to master bird portraiture. His poems and essays on nature were well received, and eight volumes of his *American Ornithology* were published between 1808 and 1814, the last of them appearing shortly after he died and one more in 1829 upon completion by his friend and biographer, George Ord. In Audubon's time the meadow vole was known as Wilson's meadow mouse (the label it was to bear in the *Quadrupeds*), for it had first been described in this country by Ord, who named it for the famous naturalist. Wilson himself had first described a number of birds, including the species that now appears in ornithological lists as the common snipe, but is known at least as well by the name Wilson's snipe.

Audubon was familiar with the published works of this indefatigable chronicler of nature, yet in spite of all this activity in the publishing of natural history (or perhaps in part because of it), he was slow to perceive the final form his art would take. Unwarranted modesty was not among his failings, but perhaps he felt he could not compete with established fame. Or perhaps he felt it would be impertinent to seek financial backing until he had recorded all possible species. He spent the years from 1810 to 1819 sketching for his own satisfaction, while attempting to support his family by other means. A second son, John Woodhouse Audubon, was born in 1812, and financial problems became increasingly severe. Still unruffled by worldly concerns, he invited Boone to tramp the woods with him again in 1813. Boone declined because of failing vision and the infirmities of age, but Audubon led parties of other hunters and sometimes used such opportunities to gather food for his family as well as specimens for his studies.

After he sold the last of his Mill Grove lands and his wife came into her inheritance, he lost everything through ill-advised ventures in fur trading, construction, steamboating, and storekeeping. There was a period when he subsisted by drawing chalk and crayon portraits for which he charged five dollars. Through these difficult times he continued to sketch and study wildlife; he was beginning to gain a minor reputation, and he maintained the almost manic exuberance of a genius possessed by an *idée fixe.* His ebullience led to pranks, one of which was to have serious repercussions. In 1818 the French-American naturalist Constantine Rafinesque visited him and battered Audubon's cherished Cremona violin one evening by using it to flail at bats that had invaded his room. On an impulse of seemingly harmless revenge, Audubon spent several days supplying his visitor with descriptions, measurements, scientific names, and drawings of ten imaginary species of fish. Rafinesque, a teacher of botany and modern languages at Transylvania University in Kentucky and a writer on a diverse range of subjects, trustingly included them in a book. Since his works were popular (if not always accurate), and since he acknowledged the source of his information, readers who detected the errors quickly realized whom he had to thank for being gulled.

Years afterward, when Audubon was struggling to achieve publication, skeptics declared that a man who had foisted nonexistent fishes on the public was probably committing another hoax by inventing some of the stranger birds that had not previously been pictured or described by Alexander Wilson.

In 1819, after a business failure involving a steam-powered lumber-and-grist mill, creditors seized Audubon's house and possessions. Jailed in Louisville as a debtor, then released on a plea of bankruptcy, he was left with nothing but the clothes he wore, his gun, and his drawings, all deemed nearly worthless by the sheriff. Again for a few months he supported his wife and sons by making chalk drawings, until friends found employment for him as a taxidermist for Dr. Daniel Drake's recently established museum in Cincinnati. Dr. Drake was generous, paying Audubon a hundred and twenty-five dollars a month, until museum admissions fell below expenses. Lucy Audubon and Dr. Drake then encouraged him in a momentous decision. His portrayals of birds, they pointed out, were better than Wilson's—more natural, more accurate, and enhanced by occasional details of appropriate habitat. Moreover, he already had pictures of some species that had been unknown to Wilson and other contemporary naturalists. Why not complete the collection and publish a comprehensive delineation of America's birds? Recognition could be his, and wealth enough to pursue his art and his study of nature for the rest of his life. He would succeed Wilson as America's foremost naturalist. To support the family while he was engaged in this enormous undertaking, Lucy volunteered to work as a governess, and she did so for seven years, beginning in 1820.

That fall Audubon set off down the Ohio and Mississippi, paying his expenses by drawing portraits while he explored the Mississippi Valley to the Gulf of Mexico. In avian numbers and varieties (and perhaps in the abundance of other wildlife at that time) the great river valley was America's richest region. His journal of observations and experiences was begun on his day of departure, October 12, 1820, when he wrote: "Without any money my talents are to be my support and my enthusiasm, my guide in my difficulties, the whole of which I am ready to exert."

He took with him an able young botanist, Joseph Mason, to whom he gave art lessons in return for help in collecting specimens. Audubon drew and described in detail every bird species he could find. In Louisiana he wandered through the bayous sketching live birds, and he bought or bartered for the dead ones kept in plumage at New Orleans markets. There he supported himself by working as a portraitist, a private art teacher (until his irritability and impatience led to dismissal), and even as a painter of street signs. Lucy brought the boys to New Orleans and then to a nearby plantation in West Feliciana Parish, where she served as a governess and mistress of a private school. Audubon moved on to Natchez, and soon afterward brought his sons there to attend school. He taught art at a local academy, an unsatisfactory situation, particularly after his field assistant Mason returned to Ohio, for it left little time to study and paint nature. He therefore came back to Beech Woods, the plantation where Lucy was employed, and became a part-time teacher of art and music. During this period he painted more than eighty of his pictures, including such celebrated and dramatic portrayals as that of a rattlesnake attacking a mockingbird nest in a tangle of yellow jessamine vines.

The pleasant, productive Beech Woods interlude ended abruptly when he refused to accommodate his employer, Mrs. Robert Percy, by brightening the cheeks in portraits he had painted of her children. He returned to Natchez, where both he and his son Victor contracted yellow fever. With Mrs. Percy's approval, Lucy Audubon brought her husband and son back to the plantation and nursed them through their illness. By now Audubon felt that his work was ready for publication, but first he wished to take Victor on one of his journeys and to find employment for the boy. In the fall of 1823, they made their way to Shippingport, Kentucky, traveling by foot for considerable portions of the journey. Audubon noted proudly that a couple of companions they gathered en route were unable to keep up the pace, while Victor, still weakened by his recent illness, outdistanced them and seemed to gain strength after days of hiking rough backwoods trails. At Shippingport Victor found employment as a clerk and his father went east, to Philadelphia and New York, in search of a publisher. Several scientists praised his work, yet no one was willing to finance its publication. The production and hand-coloring of prints was costly, and there was a general accord that Alexander Wilson's volumes, though inferior to Audubon's birds both scientifically and artistically, would overshadow the sale of any new book of natural history.

A year after his departure for the North, Audubon appeared again at Feliciana Parish. Penniless and in tatters, he had acquired no aura of defeat but rather a missionary zeal. For the sake of publication he was even prepared to forsake painting while at last making and saving money. He would then sail for Europe, there to find subscribers who would value his paean to nature in America. With his wife's help he attired himself respectably and was soon traveling from one plantation to another in Louisiana and Mississippi, charging high fees as a fashionable dancing master, violin teacher, and fencing instructor. In the spring of 1826, with his own savings and Lucy's, he took passage from New Orleans. During the long voyage, when he might easily have succumbed to fear of ultimate defeat, he self-confidently passed the time drawing creatures of the sea and air.

His assessment of his work was immediately vindicated at Edinburgh, Liverpool, Manchester, London, Paris. "The most magnificent monument yet erected to ornithology," declared the Paris Academy of Sciences. An English publisher, Robert Havell, extended credit for the production of *The Birds of America,* and issued a prospectus that quickly attracted subscribers. George IV, the profligate, unpopular, but perversely glamorous king of England, himself subscribed, thereby helping to set a fashion.

More than a decade was required to complete publication of the four hundred and thirty-five plates comprising *The Birds of America,* and during that time Audubon and his agents found a thousand subscribers each of whom was willing to pay a thousand dollars for a complete set. Audubon carefully supervised the engraving and hand-coloring of many of the plates, remaining abroad during the first three years of production and then making periodic trips with his wife between Edinburgh and America. Unfortunately, some of the subscribers failed to pay promptly, and some canceled their orders after waiting impatiently for the promised work, which was issued in parts. There were times when Havell lacked funds to pay the engravers, lithographers, and colorists, and many years passed before Audubon was able to send the last of the money he owed the publisher.

The term "elephant folio," applied to the first publication of the prints, was the British designation for a very large size of drawing paper, usually twenty-three by twenty-eight inches. The sheets in this instance were even larger—twenty-six and a half by thirty-nine and a half inches—for Audubon had drawn most of the birds life-size, which was how he wished them to be seen. Finally, in June, 1838, complete sets were ensconced in four immense morocco volumes.

Perhaps in part because they were so unwieldy and costly, Audubon and his later collaborator, John Bachman, settled on a size of twenty-eight by twenty-two inches for the even more difficult project of delineating America's quadrupeds. Most of the animal compositions were horizontal, although many of those showing tree-dwellers appropriately were vertical. The dimensions accommodated more or less life-size drawings of the smaller creatures, which were labeled "Natural Size," but there could be no consistent scale in a work encompassing some species that weighed less than an ounce and others that weighed more than a thousand pounds. The bobcat was labeled "3/4 Natural Size," the peccary "4/7 Natural Size," and so on. Unfortunately, the printer omitted these notations under some plates, and there were misprints. Most startling was the bison labeled "Natural Size," though a second bison plate bore a legend with the roughly true scale of "1/7 Natural Size" and the text correctly stated that a large bull generally weighs nearly a ton. Even in the case of the birds, the artist soon realized the futility of life-size portraiture, and a few years after completion of the elephant folio a smaller edition was published.

While Havell was producing the huge first-edition bird plates, Audubon labored to prepare an accompanying text, *Ornithological Biographies.* Though much of it was based on his own notes and journals, he worked chiefly in Edinburgh with a collaborator, William MacGillivray, a scholarly naturalist who supplied a great deal of zoological information. The first of the five companion volumes appeared in 1831, the last in 1839. The 1830s must have been a trying period, yet during his long visits home Audubon found time to investigate the woods and wildlife of such widely separated locales as Maine, Texas, and Florida.

It was on a trip from Virginia to Florida in 1831 to search for new avian species (even though his bird portrayals were already being printed) that he met his most important collaborator, John Bachman, Doctor of Divinity and pastor of the Lutheran church in Charleston, South Carolina. Audubon had arrived in Charleston with two assistants: a landscape painter and an English taxidermist. After spending the night at a boardinghouse, Audubon was looking for cheaper lodgings. As a local minister was guiding him about the town, Bachman happened by on horseback, was introduced, and insisted that Audubon and both aides stay at his home. Bachman, a devoted amateur naturalist since boyhood, was delighted to have such visitors. They planned to resume their journey after a few days' rest but stayed on for a month, observing local wildlife, collecting specimens, making notes, and discussing natural history. Whereas Audubon was the more knowledgeable ornithologist, Bachman had a more scholarly grasp of mammalian studies. And, indeed, Audubon inspired him to channel his interest into truly scientific research. When Audubon departed he and Bachman had established a friendship that was to survive financial difficulties, disagreements, and all the frustrations inevitable in an unprecedented scientific collaboration— the great work on America's quadrupeds—at a time when scientific equipment and facilities were relatively crude. Two of Bachman's daughters married Audubon's sons, and eventually several members of both families became involved in the preparation of the mammalian volumes.

In the years between the first meeting and the decision to collaborate, Bachman—though five years younger than Audubon and still devoting most of his time to the ministry—gained recognition as a scientific scholar. He corresponded with eminent naturalists and was visited by some of them. In 1838 he crossed the Atlantic to study the mammal collection at the British Museum and afterward had the honor of addressing a congress of naturalists at Freiburg, Germany.

The next year, when the *Ornithological Biographies* had just been published, Audubon was in need of funds, as always, and seeking new outlets for his restless energy. Deciding to bestow on America's mammals the same monumental effort he had lavished on birds, he impulsively printed a prospectus for publication of a work to be called *The Viviparous Quadrupeds of North America* (viviparous meaning creatures whose young are born live). He had no publisher or patron to finance the undertaking, and his animal drawings and notes were as yet scanty, but he assumed he could do all the writing himself, as well as collect, identify, and paint all the specimens. It was at this point that Bachman sent his cautionary letter, offering assistance, and Audubon deferred to common sense by inviting his friend to participate. Bachman accepted. He wanted no personal remuneration; he would work for his own satisfaction and in the hope that the endeavor would eventually help to provide for his daughters' families.

They agreed (to Bachman's distress) that they would restrict their scope to terrestrial animals, the "quadrupeds" of Audubon's prospectus. There would be no whales, walruses, or other marine mammals, nor would those winged oddities, the bats, be represented. Bachman predicted that no more than a hundred species would have to be included. Later he doubled the estimate and ultimately described more than two hundred mammals, of which a hundred and forty-seven

Among mammals first described by Bachman and Audubon was the "Little Harvest Mouse," now called the eastern harvest mouse (*Reithrodontomys humulis*). Audubon painted it nibbling corn, but he and Bachman noted its preference for weed seeds and correctly surmised that it was harmless to crops.

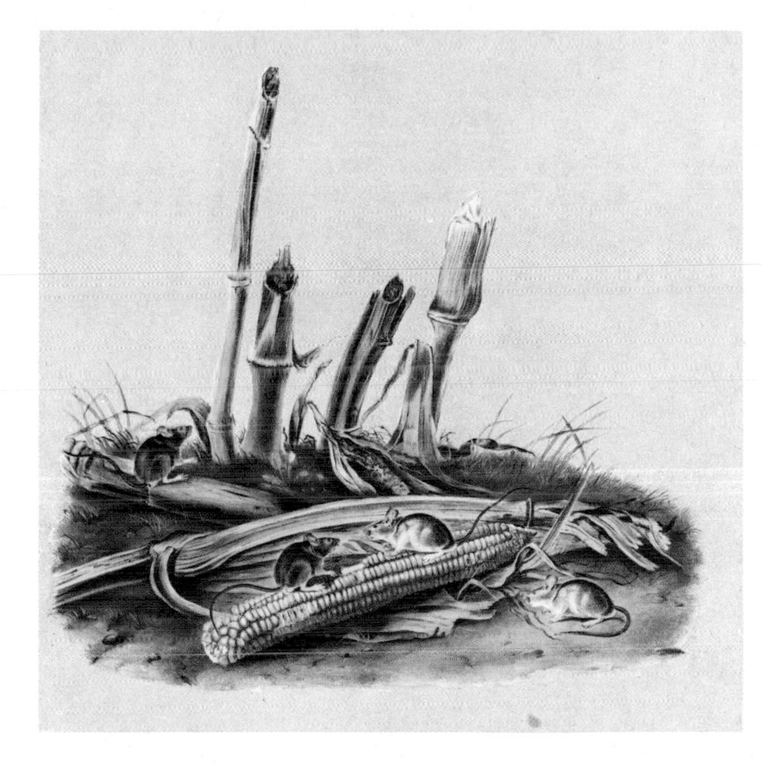

species plus eight regional "varieties" were depicted on a hundred and fifty plates. Some, however, were not really distinct species but subspecies, color phases, or even more minor genetic variations. There was an understandable tendency among naturalists of that pioneering era to divide types of animals into many subcategories on the basis of the slightest differences, "discovering" (and often naming after themselves) one new form after another. Today, with classifications founded on greater taxonomic sophistication, the trend is to group minor variations of form as single species. For example, whereas Audubon and Bachman treated two subspecies and a color phase of the eastern gray squirrel as three distinct species, each with its own Latin designation and common name, all three are now recognized as *Sciurus carolinensis*. (This volume lists the species by their modern scientific names rather than by outdated classifications.) The distinctions made by Bachman and Audubon actually represent a hundred and eighteen valid species—a far more comprehensive and detailed compilation of America's mammals than anything published before or for many years afterward.

Audubon expected the specimen collecting and painting to occupy two years. Even before recruiting Bachman he had begun painting and had written to distant correspondents requesting specimens. The methods and preservative materials available to investigative scientists of his day are revealed in his instructions to Increase S. Smith, of Hingham, Massachusetts, who was to provide varying hares: "The animals ought to be put in a Keg of Common Yankee Rum, and as soon as possible after death, cutting a slit in the abdomen of not exceeding *Two Inches* in length, and pouring Rum in the apperture until well filled."

Although a few plates were printed as early as 1842 (thus demonstrating to subscribers that the work was progressing), only about half the paintings were finished by 1846, when Audubon's powers began to fail and his younger son, John Woodhouse Audubon, assumed responsibility for completion of the animal portraits. At first the collaborators had thought the field studies and writing would take only a year longer than the painting, so that the schedule called for completion of plates in 1842 and publication of the entire opus in 1843. Audubon lived to see the last of the plates, Volume III, come from the press in 1848, but the writing was not finished until 1852, a year after he died, and the third and final volume of text was published in 1854.

Bachman's role in the undertaking was as large as Audubon's. He declined the presidency of South Carolina College partly for the sake of the *Quadrupeds*. During the twelve years of his research, correspondence, and writing, his wife died and he also lost three of his eight children, including the daughters who had married Audubon's sons. Persevering despite these calamities, he worked even when stricken by a number of ailments and suffering from a recurrent impairment of vision. He determined the validity of species collected or reported. He compiled a synonymy of species names and calculated their geographic distributions on the basis of all available publications and documents, reports from correspondents, and observations of everyone involved in the work. He wrote the basic scientific descriptions, adding to them his own observations of behavior and habits of animals with which he was familiar. He revised and edited similar accounts as well as field anecdotes and other information contributed by John James Audubon, John Woodhouse Audubon, and a number of others who volunteered or were

In addition to the plains pocket gopher (*Geomys bursarius*) that Audubon saw on the Missouri, he painted three races—mistaking them for distinct species—of northern pocket gopher (*Thomomys talpoides*). The *T. t. douglasii* variety portrayed in this plate exists only along a seven-mile stretch of the Columbia river in Washington.

asked for field observations. He then combined all this material into the final text. Some critics have charged him with diluting the elder Audubon's original and colorful style. All the same, the result was a forceful, detailed, personalized text, and an astonishingly accurate one for its time.

Several other colleagues deserve recognition for their contributions. Most important, perhaps, was John Woodhouse Audubon, a skilled painter who had made his living during the 1830s as a portraitist in England and Scotland. He joined the enterprise in 1839 as assistant artist, secretary, and general aide. In the winter of 1845-1846, when it became obvious that his father could no longer withstand the rigors of extended field trips, he made an expedition to Texas in order to observe and collect animals. Unfortunately, he was not a gifted naturalist and the expedition was not very productive. but he then spent almost a year abroad, painting American mammalian specimens (chiefly from northern Canada) at such institutions as the British Museum.

At about this time his father, though only sixty-one years old, succumbed to a progressive physical and mental deterioration. The precise nature of his ailments—the result, perhaps, of deprivation and arduous expeditions in earlier years—can only be surmised, but their effect was to dim his vision and make him incapable of sustained artistic effort.

John Woodhouse Audubon now assumed the task of completing the many animal portraits still to be painted. Unlike his father, he had had considerable academic training, and he usually worked in oils. However, to maintain uniformity in the folio, he emulated his father's methods and style, achieving such similarity that many of the pictures cannot be credited with certainty to one or the other painter. According to Bachman's text, one of two "species" (actually subspecies) of lemmings depicted in one plate was drawn in London by the son from a skin originally obtained on a tributary of the Peace River; yet the legend beneath the plate bears the name of the father. In several other instances plates bearing the father's name in the imperial folio (the first edition, entitled *The Viviparous Quadrupeds of North America*) were assigned to the son in an octavo edition (three smaller volumes, combining text and illustrations, published in 1849, 1851, and 1854 with the title shortened to *The Quadrupeds of North America*). John Woodhouse Audubon's feeling for composition was inferior to his father's, as was his mastery of the dynamism in an animal's movements, yet no one can be sure which of the two artists painted some of the finest wildlife studies in the collection. The octavo edition credited John Woodhouse Audubon with seventy-two, without counting pictures on which both men worked.

Audubon's eldest son, Victor, was an accomplished landscapist who also assisted by painting some of the backgrounds and plants, and he performed a more needed service by dealing with lithographers and printers, supervising the engraving and coloring to ensure nearly perfect fidelity to the original pictures, and preparing text for the typesetters. To reduce mounting costs and expedite printing, he condensed and further edited some of the copy sent by Bachman, and one of many difficulties arose when Bachman objected. All the same, both sons contributed substantially to the success of the project, the ardors of which overshadowed their careers and probably shortened their lives. Victor Audubon died at fifty-one in 1860, and his brother died two years later at the age of forty-nine.

Still another helpful artist—an amateur rarely mentioned in commentaries on these early renderings of wild creatures in their natural habitats—was Bachman's sister-in-law, Maria Martin. When Bachman requested pictures of insects, plants, and mammals about which he was writing, she made fine drawings for him, and also worked as editorial assistant and copyist. In 1848, two years after Bachman was widowed, he married her. Some of the others who rendered significant assistance have been singled out for recognition by the mammalogist Victor H. Cahalane in a scholarly introduction to a 1967 reprinting of plates from the imperial edition. Edward Harris, for example, contributed not only specimens and observations but funds, and, perhaps more importantly, arbitrated occasional disputes between his friends Audubon and Bachman. Spencer F. Baird, later to be secretary of the Smithsonian Institution, searched the literature for allusions to new species (those not yet classified and described by naturalists) and sent both living animals and mounted specimens and skins. Sir George Simpson, governor of the Hudson's Bay Company, supplied skins of a number of Arctic fur bearers. John K. Townsend lent the collaborators his own excellent field notes and specimens, which included hitherto undescribed species from the Rockies and the Pacific Northwest. (Just as the brush-rabbit subspecies of cottontail, among others, bears a scientific name honoring John Bachman, the white-tailed jack rabbit has a designation honoring Townsend; the former is *Sylvilagus bachmani* and was once called Bachman's hare, while the latter is *Lepus townsendii* and was once called Townsend's Rocky Mountain hare. Ironically, another cottontail subspecies was not included in the *Quadrupeds,* but has since been named Audubon's cottontail, *Sylvilagus auduboni.*)

When work was begun in 1840, Audubon was residing in New York City, though he hated its crowds and noise. After a year of feverish effort he had completed three dozen animal paintings as well as some descriptive writing. He had already found a few subscribers and was receiving at least some small profit from his work on birds. Moreover, men of means who were interested in nature were beginning to express a willingness to help finance ventures of this kind. He now acquired about thirty acres of land at the northern end of Manhattan Island, overlooking the Hudson, and by the spring of 1842 had built a house there. As the Audubon sons often called their mother Minnie, he named the property Minnie's Land. There he painted, studied, made observations of wildlife, and kept live specimens ranging from squirrels, meadow voles, and a marten to deer and elk.

As early as 1839 he had discussed the work on mammals with Robert Havell. But Havell had not yet been fully paid for *The Birds of America* and would not agree to begin another such mammoth project. Subsequently, Audubon approached J. T. Bowen, a skillful, English-born lithographer in Philadelphia. By the time Audubon moved to Minnie's Land, Bowen had agreed to lithograph and color the plates, with the actual printing to be done by other Philadelphia craftsmen. As sheets came from the press they were bound in folios of five, and a few were to be released late in 1842. By midsummer Audubon therefore felt an urgent need for specimens and information from the West, and with publication in progress he also needed funds. Hoping to obtain governmental support for an expedition, he hastily left New York for the nation's capital.

Support was not forthcoming, though in a manner of speaking it was offered posthumously in 1857, when Congress appropriated $16,000 for the purchase of one hundred sets each of the *Birds* and the *Quadrupeds* "to be presented to foreign governments in return for valuable gifts made to the United States." Despite governmental rejection the trip to Washington was a success, for at his hotel Audubon met Colonel Pierre Chouteau, a fur trader and financier who was to become a major supporter. In addition to providing skins of western animals, he enabled Audubon, together with his friend and fellow naturalist Edward Harris and three modestly salaried assistants, to make the western expedition. Chouteau was connected with the American Fur Company, which operated a steamboat on the upper Missouri. Harris rented his farm to meet a fifth of the expedition costs, and in the spring of 1843 the Audubon party took the steamboat to the Fort Union trading post, situated on the Missouri near the confluence of the Yellowstone. Headquartered there from June 12 to August 16, the five observed and collected a large assortment of mammals, including such diverse species as prairie dogs weighing a couple of pounds, pocket gophers less than half as large, and bison weighing a ton. In excitement and curiosity, Audubon approached one of the big bulls and narrowly escaped when the animal charged.

It was here, on the day he arrived, that he saw his first bighorn sheep. In fact, he saw twenty-two of them—a band of ewes, one lamb, and some rams (presumably young rams, since, as Audubon perceptively noted, mature males form bachelor groups except at breeding time). He wrote that "they scampered up and down the hills much in the manner of common sheep" and were so sure-footed that they also bounded up and down the steepest peaks and ravines. He knew that where sheep had never been hunted they sometimes permitted approach, but here they had learned to fear man, and he and his companions had to augment their observations with specimens provided by others. "Notwithstanding all our anxious efforts to get within gun-shot . . . we were obliged to content ourselves with this first sight of the Rocky Mountain Ram." Still, he produced an impressive painting of a ram and ewe representing the now- extinct Badlands race of bighorn sheep. The part of eastern Montana he explored was not far from the fringes of the Badlands, and he mentioned that the French Canadians called the country *mauvaise terres.* He insisted that he found writing tedious, but his description of sheep habitat and behavior in these mountains typified his concise evocations of nature:

"In many places columns or piles of clay, or hardened earth, are to be seen, eight or ten feet above the adjacent surface, covered or coped with a slaty flat rock, thus resembling gigantic toad stools, and upon these singular places the big horns are frequently seen, gazing at the hunter who is winding about far below, looking like so many statues on their elevated pedestals. One cannot imagine how these animals reach these curious places, especially with their young . . . which are sometimes brought forth on these inaccessible points, beyond the reach of their greatest enemies, the wolves."

His journal described landscapes, plants, animals, trappers, hunters, Indians. He made many drawings, collected specimens, and brought back a live badger, deer, and the diminutive buff-yellow breed of reynard known as the swift fox. But for all his recording and collecting, he knew when he reached home in autumn that the expedition had not been entirely successful, and he felt so old and tired that he realized he would never make another wilderness journey. John Bachman had expected a vast store of western information; he was disappointed to receive only colorful generalities and anecdotes about some of the unfamiliar species in place of precise scientific data. Audubon declared apologetically that besides his many sketches he had returned with specimens which might well prove to represent some new birds and a new variety of pronghorn antelope. Attempting diplomacy, Bachman praised the birds but warned that the pronghorn had probably been described in previous works; and so it had. (Audubon did, however, prove to less diligent observers that this unique American animal sheds only the sheaths of its horns—not the whole structure as antlered species do, and unlike other horned species which shed not at all.)

Eagerness to identify new species led to frequent errors by naturalists. In the same year, 1843, Audubon described an odd bird specimen that had been shot a couple of decades before in Louisiana. He named it the Brewer's duck (after Thomas M. Brewer, an ornithologist he esteemed), then added a candid notation that it might not be a valid species but merely a hybrid of the mallard and some other familiar duck, "perhaps the Gadwall, to which also it bears a great resemblance." He was normally more cautious in such matters than many of his contemporaries and in this case his suspicion was correct. Had he been equally skeptical and thorough at Fort Union, the subsequent compilation of text would have been easier.

Bachman felt that Audubon, entranced by the impressive western game animals, had spent too much time watching bison, wolves, grizzlies, and those unique pronghorns and too little observing and collecting skunks, rabbits, hares in winter pelage, and the like. At his insistence, the artist requested additional specimens from Fort Union, but these did not remedy what Bachman considered to be the inadequacies of the written information Audubon had sent him. In 1845 Bachman therefore visited Minnie's Land for a consultation and a look at his friend's notes, which Audubon refused to show until Bachman finally threatened to withdraw from the endeavor. The visit must have been sorrowful for Bachman, who saw that Audubon was swiftly declining. Now he understood the paltriness of the latest notes, as well as errors in the manuscript pages Audubon was sending him.

The first volume of plates, containing fifty pictures, appeared that year, followed in 1846 by the first volume of text, and evidently it was the encouragement of seeing the work published that determined Bachman to go on. John Woodhouse Audubon was now doing all the painting, while Bachman continued with the text almost unaided. His vision became so dim that he dictated the final portions to Victor Audubon, but he nevertheless completed the third volume—the end of the great opus—early in the spring of 1852. Audubon had known, before he died in January of the previous year, that his friend would finish the task, and the knowledge must have brought him a measure of contentment. Fortunately, Bachman lived until 1874 and saw the work receive some of the praise it merited before he died at the age of eighty-four.

The praise has periodically subsided and swelled. Occasionally, critics have objected that some of the animals were stiffly portrayed, anatomically inaccurate, or in contorted postures. A few of them were, certainly. But they were the most natural animal pictures ever published at that time, and the inspiration for later masters of nature like Ernest Thompson Seton, Louis Agassiz Fuertes, and Roger Tory Peterson. Many of the paintings were also exquisitely stylized and composed. Moreover, they were a significant departure from typical earlier scientific illustration in that they revealed animals in motion, lively, dynamic, and amid their natural surroundings of cover and food, rather than as dull, dead, old-fashioned museum exhibits. Some, perhaps influenced by the work of artists like Peter Rindisbacher and Thomas Hewes Hinkley, were narratives. Audubon's coyotes greet each other at a water hole, his black bears are about to contest the remains of a deer kill. Thus he painted more excitingly than other illustrators of nature and also helped to prepare the way for the ecological view of wildlife.

Surprisingly few books have reprinted even his best-known animal plates, and until recently no process of reproduction could adequately convey the quality of the originals. Some of the artistic objections, in fact, seem to have been based on faulty reproduction. For example, in the only two books devoted in recent years to Audubon's quadrupeds, the gray fox assumes an inexplicably strange stance—walking in a crouch with its nose pointing skyward as if some invisible heavenly presence had arrested its attention in midstride. The position appears excessively stylized since there is no motive for the sharp uptilting of the head. This is because crucial detail was lost in the printing. Such improvements have been made in lithography that now Audubon's gray fox, Plate 21, can be reproduced faithfully in all its detail—including a feather fluttering above the animal's nose. It is this light, delicate breast feather, perhaps from a flushed duck or goose, perhaps from the poultry raised at the farm in the background, that has caught the attention of the hungry fox. With the feather visible above its raised snout, the position no longer seems unnatural but dramatically taut.

Similarly, the colors in reproductions have often been unfaithful, owing to faded or damaged original sets or less than ideal engraving and printing. This, too, has been remedied and it is possible to see how brilliantly, as well as accurately in most cases, the animals and foliage were rendered. No printed depictions of animals equaled the realism of their hand-coloring until well into the twentieth century.

Another criticism concerned the haphazard order of plates in both of Audubon's great compilations. The first quadruped presented was a bobcat, the second a woodchuck, the third a white-tailed jack rabbit, and the hundred and twelfth a black-tailed jack rabbit; a white-tailed deer was preceded by a water vole and followed by a sea otter, and so on. This was unavoidable. Audubon painted the animals as he encountered them or received specimens. He sometimes put aside small, easily preserved specimens to work on large ones while they remained fresh. And as pictures were completed they had to be rushed to the engravers. Had he waited to put the species in an order of families or natural relationships—all rodents together, all carnivores or ungulates—the work could not have been completed. When his plates are reprinted today, they are usually rearranged in taxonomic sequence, which surely would have been Audubon's method had he been free to choose.

One can most fully appreciate the sharpness of the artist's perceptions by viewing the finest plates in conjunction with photographic studies of the animals made by the most painstaking and sensitive wildlife photographers at work today. In this way each creature is seen not just in a single prototypical pose, but in the full, rich context of its behavior and habits, perhaps with its mate, its young, its competitors, companions, or enemies, its shelter and foods—the biome that gives it life and brings to life the depth and scope of Audubon's vision.

No. 14.
Drawn from Nature by J.J. Audubon, F.R.S.F.L.S.
DIDELPHIS VIRGINIANA, PENNANT.
VIRGINIAN OPOSSUM.

PLATE

66

Opossum

Didelphis marsupialis

"The whole structure . . . is admirably adapted to the wants of a sluggish animal," reported Audubon and his collaborator, John Bachman. Its long jaws hold "a greater number and variety of teeth than any other of our animals, evidencing its omnivorous habits; its fore-paws . . . aid in seizing prey. . . . The construction of the hind foot . . . and its long nailless opposing thumb, enable it to use these feet as hands, and the prehensile tail aids it in holding on to the limbs of trees." The account was perceptive and the painting showed opossums harvesting their favorite food, ripe persimmons. This primitive species, North America's only marsupial, inspired absurd folklore as well as a belief by many early naturalists that the young were produced by "mammary gestation," sprouting from the pouched nipples. Audubon and Bachman, rejecting myths, discovered that the tiny, hairless infants, hardly more than embryos, use their forelegs and hooked claws to creep into the pouch, and emerge a month or two later looking like shaggy mice. As Audubon noted, they ride the mother's back while grasping her arched tail with their own. Even in his time these unique creatures probably were extending their range northward. They have now reached Canada.

An oppossum may bear a dozen and a half young, but infant mortality is high. By early summer the few survivors are big and active enough to leave the mother's pouch and ride on her back. In another month they will disperse. "Playing 'possum"— curling up, as at left, and feigning death—is an occasional and involuntary defense mechanism. It is demonstrated by diverse creatures, including more than one species of snake. A disturbed opossum more often seeks a hole or climbs a tree. If cornered it may hiss and salivate, baring all of its fifty teeth.

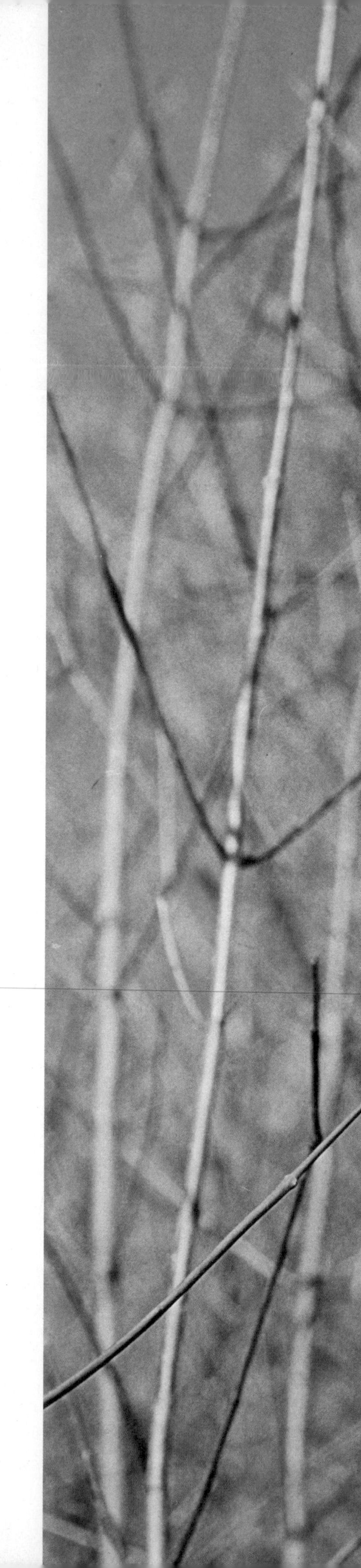

PLATE

146
Armadillo
Dasypus novemcinctus

The armadillo pictured is the nine-banded variety, North America's only species. This peculiar mammal is more than two feet long, including the tail—"the size of a large opossum," as Audubon and Bachman said. One of the more tropical species is a bit longer, while another is mole-size. A relative of anteaters and sloths, it feeds on insects, with an occasional garnish of vegetation. It has grinding teeth but no canines or incisors, and since its claws are better for burrowing than fighting, the bony armor plates are its chief defense. Withdrawing its head almost like a turtle and flexing the jointed bands, it curls to shield its soft belly. However, Bachman erred in believing that this species "rolls itself up" into as compact a ball as some South American armadillos. The four young in a usual litter are all of one sex, having developed from a single divided egg. As their skin is soft, ossifying slowly, they resemble a flock of miniature piglets trailing their mother to snuff and root for ants on a southwestern desert or prairie.

Nº 30.

PLATE CXLVI
DASYPUS PEBA, DESM.
NINE-BANDED ARMADILLO.

The North American armadillo usually forages near one of its several burrows in an area of less than ten acres. It is known as the nine-banded species but is not invariably girdled by that many scutes. Two of the animals pictured here have eight rings, and an occasional specimen has only seven. Some mammalogists have reported that it cannot roll into a defensive ball, but the photograph below proves that it does—though not as tightly as some of the twenty Latin American species. Another peculiar ability is that of walking underwater to cross streams, but it can also swim—gaining buoyancy by gulping air. It is common from Mexico to Kansas and seems to be spreading its range eastward.

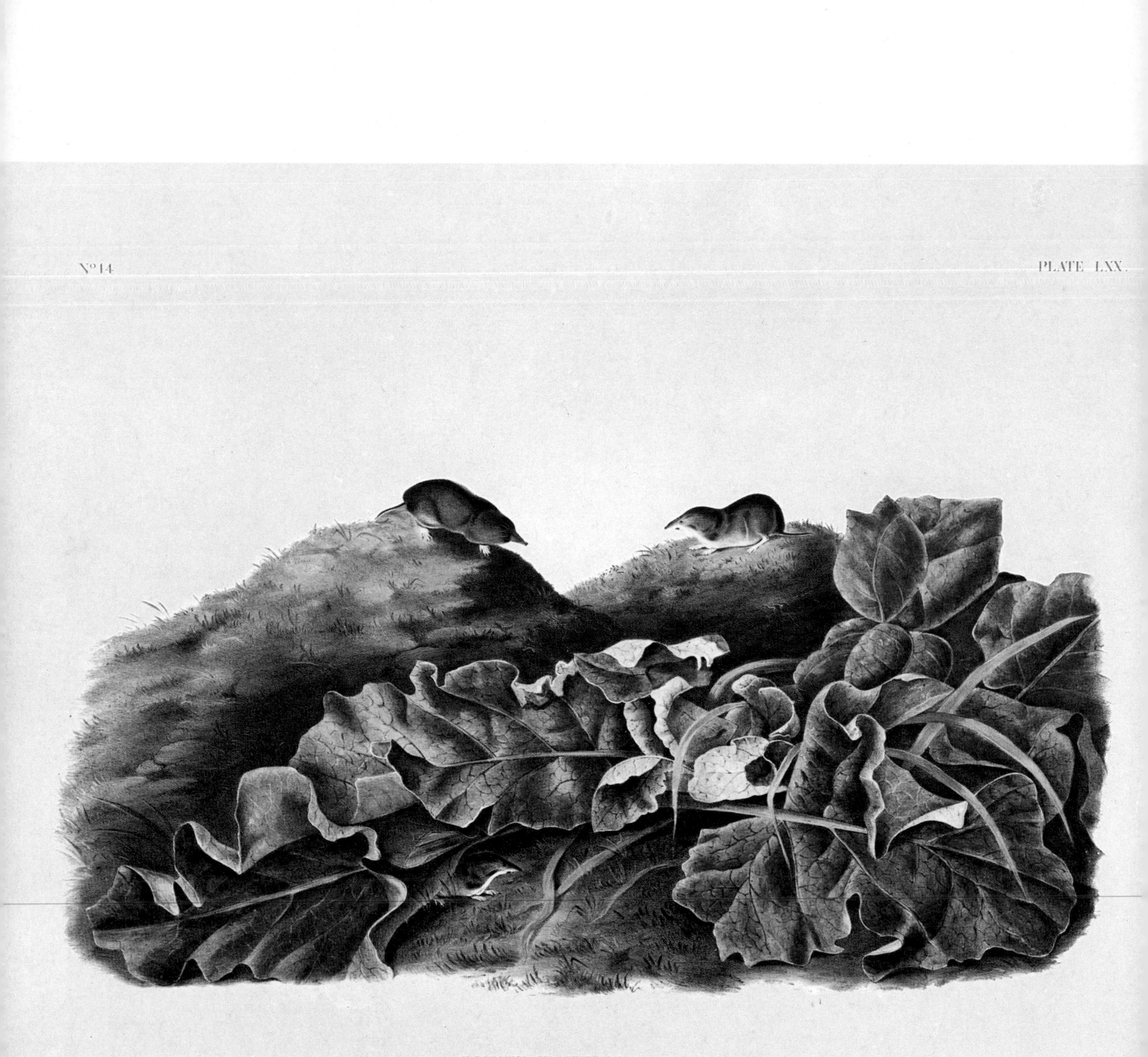
N°14.
PLATE LXX.
Drawn from Nature by J. J. Audubon, F.R.S. F.L.S.
SOREX PARVUS, SAY.
SAYS LEAST SHREW.
Natural Size
Lith. Printed & Col. by J. T. Bowen, Philad. 1845

PLATE

70
Least Shrew
Cryptotis parva

As high-strung as their human namesakes, shrews belong to the order *Insectivora,* most primitive of mammals and probably close to the ancestral stock of all mammalian forms. The least shrew vies with the pigmy shrew as the world's smallest mammal. The pigmy, half as heavy as a dime, has less body but more tail, and either species is unlikely to be more than three inches long from the tip of that tail to the point of its snout. The metabolism of the velvety, cinnamon-tinged least shrew demands voracious feeding. The animal probes loose soil or leaf litter for insects, worms, snails, salamanders, even small frogs. Its tiny eyes and ears are ineffective, its sense of smell weak, but its sensitive vibrissae—stiff hairs growing around the mouth—detect prey. It is, in turn, occasionally snatched up by larger carnivores despite an unsavory musk exuded by belly and flank glands. Another species, the short-tailed shrew, paralyzes prey with venomous saliva but is not dangerous to man. A dozen or so American varieties are widely distributed, the least shrew inhabiting most of the East. Audubon wrote that a specimen was first collected for science in a Nebraska pitfall meant for wolves. Captured by the animal painter Titian Peale, it was named by Thomas Say, an entomologist, during an expedition to the Rockies.

Only a trifle heavier than a pygmy shrew *(Microsorex hoyi)*, the least shrew is apt to weigh no more than a fifth of an ounce. The penny in the portrait at left vividly indicates its size. Though it lacks the venomous saliva of the slightly larger short-tailed shrew *(Blarina brevicauda)*, it is ferocious enough to vanquish small lizards. Another diminutive species, the wandering, or vagrant, shrew *(Sorex vagrans)* is pictured as it pounces on a grasshopper, and again with its tiny young.

PLATE

10

Common Mole

Scalopus aquaticus

To Audubon and his contemporaries most moles were known as "shrew moles," a term now reserved for a single mole species, although shrews belong to the same order of insectivores. The common, or eastern, mole is most widespread, ranging across half the continent. As with most moles, its eyes are vestigial light-sensitive organs beneath the skin and, though it hears well, its tiny ear holes are hidden by fur. Its soft, grainless coat offers no resistance as the animal moves forward or back in cramped foraging burrows. Its toes are slightly webbed and the soles of its forefeet turned out so that it can dig with a swift breast stroke, swimming through soft earth. It is hated for the ridges its tunnels raise across crop fields and manicured lawns, the mole hills formed by subterranean dens, and its habit of nibbling tubers. Audubon and Bachman, who defended some unpopular species, remarked in this case that moles eat hordes of destructive insects. Ironically, they were only partly right, as they also praised the common mole for devouring earthworms, then regarded as harmful rather than beneficial.

PLATE X.

Lith. Printed & Col.d by J. T. Bowen, Phila. 1843.

SCALLOPS AQUATICUS, LINN.

COMMON AMERICAN SHREW MOLE

Natural Size

MALE AND FEMALE

As a mole erupts from the ground
or digs in again, it seems to be swimming through
the earth, its grotesquely enlarged
forepaws turned out to part the loosened soil.
Almost equally grotesque is the
common mole's long, naked, flexible snout, which
aids in subterranean hunting by its
delicate senses of touch and smell. A species of
another genus, the star-nosed mole
(Condylura cristata) has an even stranger snout,
with twenty-two short, fleshy barblets
surrounding the nostrils. The body of every mole
species is a stout cylinder that slides
along easily in the search for grubs and earthworms.
In porous soil, tunnels can be opened
or lengthened at the rate of a foot per minute.

No. 5.
Drawn from Nature by J.J. Audubon, F.R.S. F.L.S.
LEPUS SYLVATICUS, BACHMAN
GREY RABBIT.
Natural Size
OLD & YOUNG.

PLATE.XXII.

PLATE

22

Common Cottontail

Sylvilagus floridanus

The cottontail is a meek but prolific genus that has all but inherited the unpaved earth from ocean to ocean and from lower Canada to South America. There are more than a dozen species, nearly seventy subspecies, and they are America's only rabbits. Until almost two decades after Audubon's death, scientists failed to distinguish them from hares, which are larger (though the females of both groups outweigh males, a trait unusual in mammals) and have longer gestation periods and more precocial young. The common, or eastern, cottontail was known as the "grey rabbit," but the rest were called hares. Errors were inevitable. Audubon twice pictured the Rocky Mountain, or Nuttall's, cottontail. One version, based on an immature specimen supplied by the naturalist Thomas Nuttall, was described as diminutive and labeled "Nuttall's Hare." The other, based on observations of the same species in sagebrush—widely called wormwood—was descriptively labeled "Worm Wood Hare." Commenting on the fecundity of eastern cottontails, Bachman wrote that "nature seems thus to have made a wise provision for the preservation of the species, since no animal is more defenceless or possesses more numerous enemies."

Whether screened amid summer's grassroots jungles or nestled under a browning cornstalk on snow, a rabbit escapes notice by sitting perfectly still, not even blinking. When an enemy comes too close, it can bound away at twenty miles an hour. Cottontails thrive on diverse foods, from grasses and sedges to farm crops, weeds, and brambles. If no predators, accidents, or diseases counterbalanced their multiplication, a pair and their descendants could produce three hundred and fifty thousand cottontails within five years.

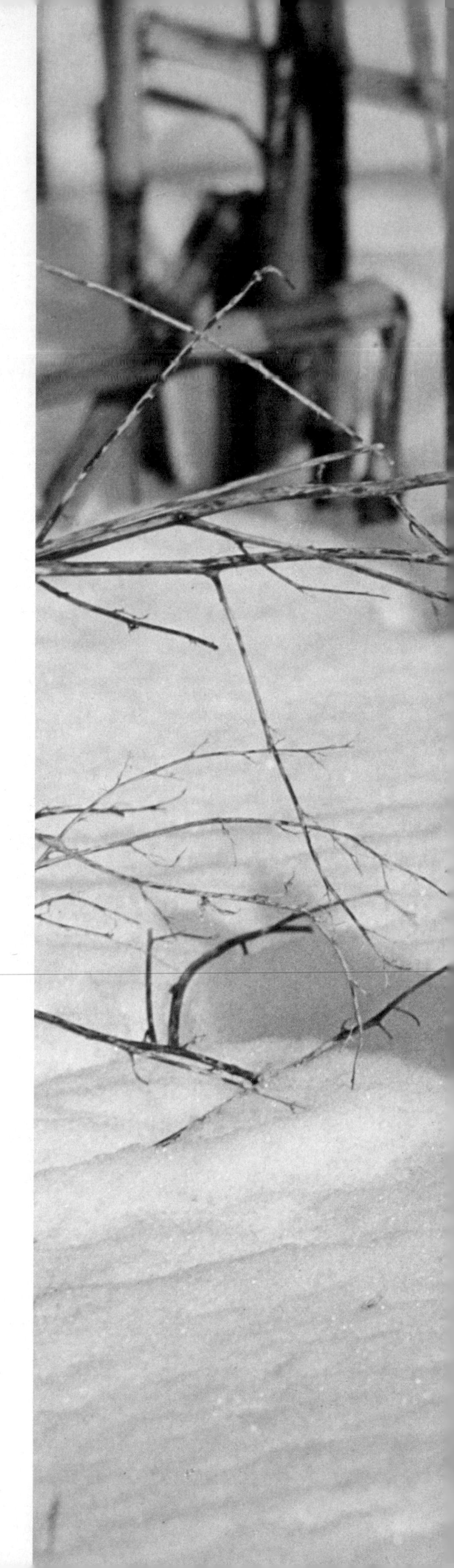

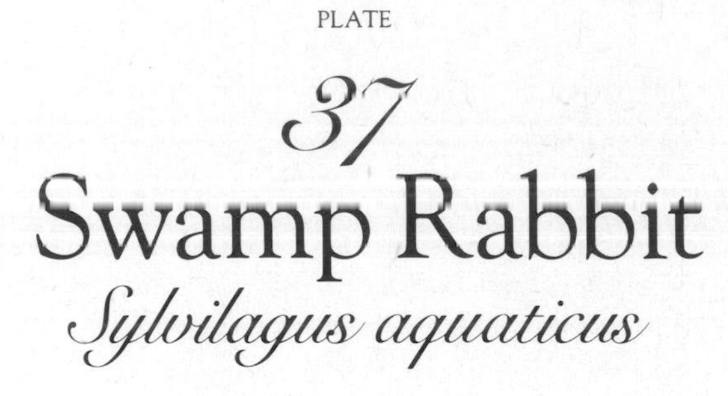

Its cottontail genus notwithstanding, the swamp rabbit has several oddly harelike characteristics. Whereas the four to six young in an average cottontail litter are born blind and naked after a month's gestation, two or three swamp rabbits are born furred, after a gestation of almost six weeks, and their eyes open almost immediately. At maturity they often weigh over five pounds—enough to pass for adult hares. Denizens of the Mississippi Basin's moist southern lowlands, they have come to be called "cane-cutters" because of their appetite for cane. They also eat sedge, grass, and aquatic herbs. Audubon and Bachman donated a specimen to Philadelphia's Academy of Natural Sciences, and the text for the *Quadrupeds* told how "this individual, on being pursued by hounds, swam twice across the Alabama river, and was not captured till it had finally retreated to a hollow tree." In its powerful swimming and choice of hiding place, it exhibited typical behavior. Splay-toed hind feet aid in paddling, and the species can nearly submerge to elude predators. A cottontail takes to water if forced to; a swamp rabbit, Bachman observed, "plunges fearlessly into it and finds it a congenial element."

PLATE XXXVII.

LEPUS AQUATICUS, BACH.

SWAMP HARE.

Natural Size.

MALE.

Lith. Printed & Col.d by J. T. Bowen, Phila. 1844.

At a distance, a swamp rabbit resembles the
common cottontail and the closely related marsh rabbit
(S. palustris), and all three may share a
southern area of low bogs and cottontail uplands. However,
the swamp rabbit is larger than either of the
others and darker than the cottontail. Only a small, young
swamp rabbit—one whose underparts retain
the gray of infancy—truly looks like a marsh rabbit.
As Bachman observed, the underside of a
marsh rabbit's tail is ashen, mixed with brown.
That of an adult swamp rabbit is the flash of white visible
in the picture at right, as snowy as a cottontail's.
Audubon's painting is lifelike in its proportions and hues
but seems to endow the creature with fully
webbed hind feet. This water-loving rabbit has between
its toes a larger membrane than a cottontail's,
but Audubon's webbing is hyperbole.

PLATE

12

Varying Hare

Lepus americanus

Audubon portrayed this northern hare, as the species used to be called, in almost full winter pelage, with its legs just turning summer brown and some of the thick fur shed from between the hind toes. It is more normal in fall for the legs and ears to whiten before the body does and remain so until after the head and body turn dusky or chocolaty in spring. Another of the naturalist's paintings showed the summer coat, the seasonal change of camouflage implicit in the name "varying hare." The popular names "snowshoe hare" and taxonomically improper "snowshoe rabbit" allude to the massive hind feet, broadening toward the front and copiously haired to add support on soft snow, prevent wet snow from sticking, give traction on ice and insulation from cold, and minimize the lingering trail scent that betrays the hapless to a lynx, bobcat, or fox. The overall insulation of fur led Bachman to contrast the animal's "shaggy" winter appearance with the sleekness of other hares and rabbits. After the snows come, varying hares browse on bark and twigs, preferably in dense, swampy conifer thickets that afford concealment. In summer they prefer the drier uplands, and Bachman noticed that after crossing swamps "they are for hours employed in rubbing and drying their paws."

No. 3.

LEPUS AMERICANUS, ERXLEBEIN.
NORTHERN HARE.
Natural Size.

Varying hares are adapted to a subarctic
climate but also exist in arctic and temperate zones—
throughout Alaska, Canada, New England, the
upper Midwest, the Appalachians
down to West Virginia, the Rockies to New Mexico,
and the Pacific chains to California.
Hares in the Cascades and Sierras do not vary but
remain brown all year. Elsewhere their
camouflage sometimes betrays them. At right, an
Alaskan specimen, stretching to nibble
twigs in September, is still so brown that its coat
will contrast with the snow if it leaves
the thickets. Two of those above are in winter
pelage. The third, undergoing the spring transition,
demonstrates a habit Bachman described of
diligently grooming its hind feet after crossing
marshy ground. The size and shape of
its paws vivify the popular name "snowshoe hare."

PLATE

3
White-tailed Jack Rabbit
Lepus townsendii

All the West is blanketed by the overlapping ranges of several big, lanky, fleet, long-legged hares named, like mule deer, for their ears. Bachman confirmed that at least two species "received from the Texans, and from our troops in the Mexican war, the name of Jackass rabbit." Alone among these hares, the white-tailed race turns pale in winter, almost as white in northern climes as a varying hare. The naturalist John Kirk Townsend, who first described this species, encountered it as a staple meat of Indians and explorers in 1836 at Fort Walla-Walla, Washington, and it erased his epicurean memories of cottontail: "Regaled with a

PLATE. III.

dish of hares . . . I thought I had never eaten anything more delicious." Recording behavioral traits, John Woodhouse Audubon speculated that another variety, the black-tailed *L. californicus*, "appears to possess just brains enough to make him the greatest coward . . . quite as wild as a deer, and equally heedless as to the course he takes." Vulnerable creatures of exposed terrain, jack rabbits—all of them—are really far from heedless. When fleeing, they make frequent high jumps to observe their pursuers and their surroundings.

Titian Peale described their broken run as "three short and one long leap."

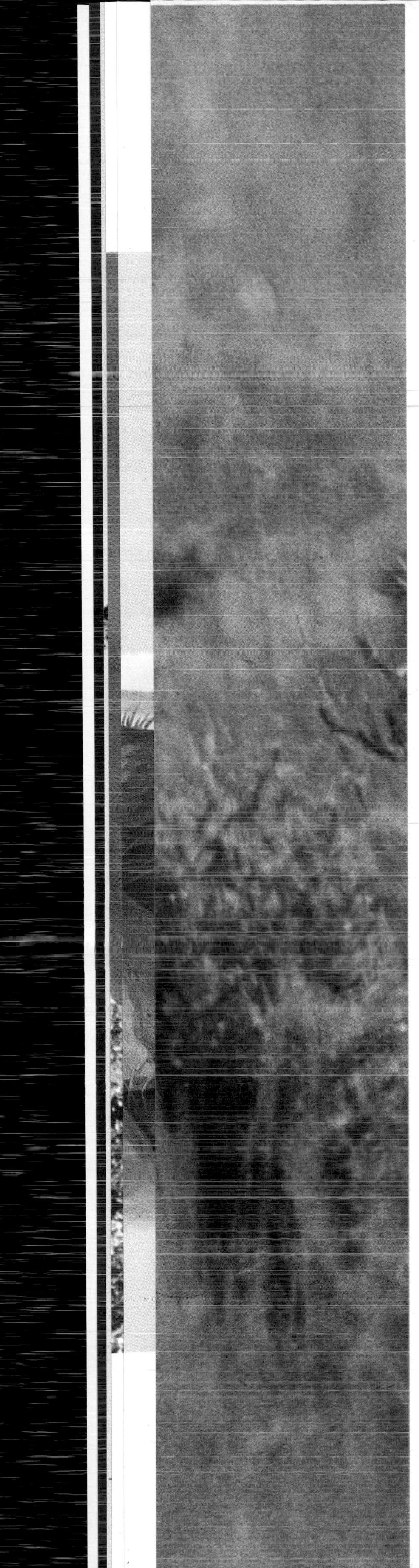

Peering over lush summer grass is a white-tailed jack rabbit. Another is seen below, clad for winter when, in spite of its long ears and legs, it is sometimes mistaken for a varying hare or an arctic hare *(L. arcticus)*. It flourishes on plains and open slopes from central Canada to New Mexico and from the Sierras to Wisconsin. The more adaptable black-tailed jack rabbit, shown in the other two pictures, inhabits deserts as well as farmlands and prairies from Missouri northwestward to Oregon and down into Mexico. It is smaller and leaner than its white-tailed relative but has even more impressive ears.

Constantly alert, spirited, and
nervous, a red squirrel perches on a stump or
a high tree limb and scolds almost
every passing creature. Anything from an
equally loquacious little nuthatch
to a silently browsing deer or a strolling
woodsman can elicit a frenzy of
chirping, tail-flicking, and even spasmodic
gestures of the forepaws as
the squirrel sits up on its haunches.
Like the Canada jay, the red squirrel is an
opportunistic snatcher of foods. All
squirrels love corn and will visit
wildlife feeding stations, as at right, to
eat whatever is offered. Pictured
above is a female in early summer, playing
with two of her four or five
half-grown young. Tree cavities and tightly
woven leaf nests are common, but
grasses or conifer needles may also be used.

PLATE

17
Fox Squirrel
Sciurus niger

Cypress swamps or mangroves may shelter fox squirrels, yet in most regions the species seeks higher, drier, sparser nut woods than the thickly timbered havens of the smaller gray squirrel. It formerly resided only in open midwestern forests edging the prairie. By the time of Audubon's observations, homesteading had cleared a new range for the animal through most of the East. Audubon and Bachman discovered what they thought to be four species and included them with another in the *Quadrupeds.* Later Audubon's younger son added a sixth. Actually, they all represented three of the ten races of one species. The northeastern form shown here was labeled "Cat Squirrel," a name probably derived from its catlike actions and some of its calls. All forms have luxuriant brushes and a vulpine lope on the ground, but only one was then called a "fox squirrel"; it was usually melanistic or washed with fox red. Confusion was inevitable, for the fox squirrel exhibits a perplexing diversity in size, color, and even skull shape. A typical southern specimen is black, perhaps with a white tail tip and face blaze. In South Carolina it is black with white ears and nose, in Illinois red-bellied with a bright rusty wash above, and in Pennsylvania more subdued, like one of these "cat squirrels," with an orange-tinted coat of pepper.

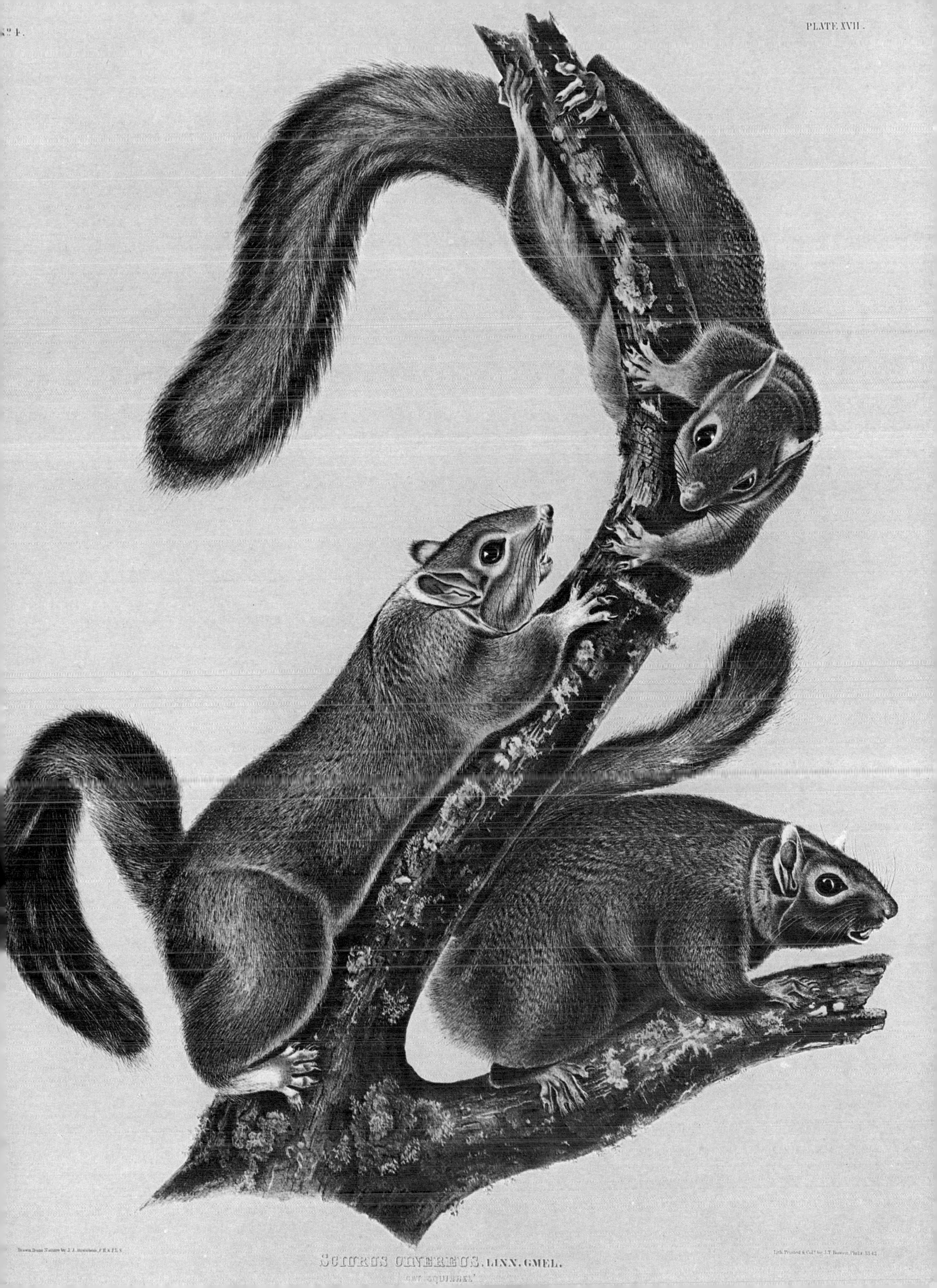

Drawn from Nature by J. J. Audubon, F.R.S. F.L.S.

Lith. Printed & Col.d by J. T. Bowen, Philada. 1843.

SCIURUS CINEREUS, LINN. GMEL.

CAT SQUIRREL.

Basking in the sun fifty feet up
in a hardwood, the fox squirrel at left resembles one
of Audubon's painted studies in Plate 17,
both in its rusted gray color and in the way it hugs a
limb. The squirrel with the whitewashed ears
and nose typifies those in parts of the
Deep South where the squirrels have adapted to lowland
habitat. It was photographed in Florida's
Big Cypress Swamp. Fox squirrels establish feeding
perches. The one on the fence rail is
carrying a nut—much lighter cargo than the corn cobs
that midwestern fox squirrels often haul
to a fence or stump from nearby fields. By the time of
the nut harvest, the three or four infants
in an early spring litter are nearly grown; their
mother has mated again and the young
of her second litter, brought forth in the summer,
are already weaned and self-sufficient.

N°7.
PLATE XXXV.

PLATE

35

Eastern Gray Squirrel

Sciurus carolinensis

"It sallies forth with the sun," Bachman wrote, ". . . in search of food . . . scratching among leaves, running over fallen logs . . . making almost incredible leaps from the higher branches of one tree to another." Acorns, beech and hickory nuts, pecans and walnuts are its delights, but the gray squirrel is an adventurous forager. In this plate a northeastern subspecies festoons a tulip tree, and Audubon painted what he supposed to be two other species aptly poised on chestnut and butternut limbs. One, however, was a southeastern race of the same squirrel, the second a common melanistic color phase. The form reproduced here was labeled "Migratory Squirrel," in reference to a phenomenon no longer seen now that the hardwoods have been thinned. Eastern gray squirrels remain abundant from lower Canada to the Gulf and across half the United States, but nowhere thrive so marvelously that overcrowding induces the mass exodus Bachman described: "This species of squirrel has occasionally excited the wonder of the populace by its wandering habits . . . and long migrations. Like the lemming . . . it is stimulated either by scarcity of food, or by some other inexplicable instinct. . . . Mountains, cleared fields, the narrow bays . . . or our broad rivers, present no unconquerable impediments."

In its life cycle the eastern
gray squirrel is much like the fox squirrel,
but it tends to be smaller, more uniformly peppery,
more agile and arboreal, and it prefers
more heavily wooded habitat. It is nearly domesticated
in city parks, shielded from most
predators while rummaging through a dole of assorted
foodstuffs. In the wild it seeks
hardwoods such as nut-laden oaks and hickories in
bottomland thickets or mature forests
with an understory of smaller trees—timber so dense
that it can travel through the crowns
whenever danger is detected below. It caches more
nuts than do other squirrels, and though
it sits up on a limb or stump to nibble morsels held
between its forepaws, it establishes no
regular feeding perches. It can descend trees
head-first with astounding speed or hang by its hind
claws to rest and scan its surroundings.

PLATE

43
Western Gray Squirrel
Sciurus griseus

Working from specimens and notes sent by distant colleagues, Audubon understandably portrayed western gray squirrels on a tree that should have attracted them but, unfortunately, did not grow in their woods—in Washington, Oregon, and California. The authors of the *Quadrupeds* wrote candidly of this species: "We know nothing of its habits, as it was brought from California, without any other information than that of its locality. We have represented two of these Squirrels in our plate, on a branch of hickory, with a bunch of nearly ripe nuts attached." A more accurate setting would have been amid western oaks or pines. The eastern and western gray squirrels are distinct species, unable to hybridize, yet if their ranges overlapped they would be hard to tell apart at a distance. The western type has a lighter belly and is slightly larger, with a narrower tail about a foot long and a body of roughly the same length. Like the fox squirrel, it does much of its foraging on the ground. Owing to the mild Pacific winters, it buries little of its food, but one of Audubon's correspondents in California observed how it sometimes stores nuts for convenient future consumption by poking them into the holes drilled in pine trees by the acorn woodpecker.

Drawn from Nature by J. J. Audubon, F.R.S. F.L.S.

Lith. Printed & Col.d by J. T. Bowen, Philad.a 1844.

SCIURUS LEPORINUS, AUD & BACH.

HARE SQUIRREL.

Natural Size

Like other members
of its clan, the western gray
squirrel arches its tail
up over its back when it pauses
to crack husks or eat
nuts, fruits, buds, and seeds.
This posture aids in
balance while keeping the tail
out of the way. At
other times the bannerlike plume
serves more critical
purposes. It becomes a rudder
and parachute during
swift climbs, leaps, or falls.
It also provides
a misleading target for diving
hawks and swooping owls,
sometimes causing them to strike
mere fur and miss the body.

No. 2. PLATE VIII.

Drawn from Nature by J. J. Audubon F.R.S.F.L.S.

Lith. Printed & Col.d by J. T. Bowen Philad.a

TAMIAS LYSTERI, RAY.

CHIPPING SQUIRREL, HACKEE &c.

Natural Size.

MALE, FEMALE AND YOUNG FIRST AUTUMN.

PLATE

8

Eastern Chipmunk

Tamias striatus

The philosophical naturalist John Burroughs, amused by bustling chipmunks on a farm in the Catskills, where he settled in 1874, watched one closely to determine how much provender it carried in its resilient cheek pouches. In three days it deposited a bushel of chestnuts, hickories, and husked corn in its tunnel and nest chamber. Bachman, too, wrote admiringly of "chops distended" by nuts and berries. *Chitmunk* is an Algonquian word for squirrel. The name "chipping squirrel" was more common in Bachman's era because, as he remarked, "its clucking resembles the chip, chip, chip, of a young chicken." When stationed near its den, this small rodent may indulge in loud bravado until a weasel or other enemy sends it dashing for cover with its tail up. In portions of the Southeast no chipmunks rustle the fallen leaves; elsewhere one species ranges from the Atlantic to the Midwest and sixteen, belonging to nearly sixty geographical races, are sprinkled from the Great Lakes to the Pacific. All have five black and four pale stripes from shoulder to rump, but few are as vividly marked or plump as the eastern form. In northern woods it hibernates. Where the winter is mild, however, it often peeps from one of the tunnel entrances strategically located beneath stumps, logs, and jumbled rocks at the edges of timber.

The three chipmunks, inquisitively
peering over a log beneath sheltering pine
boughs, are a bit more than a month old. Probably born
in April or May, after a month's gestation,
they may well have had at least one or two more siblings,
and in warm southern habitat a
second litter often arrives in July or August.
At birth, a chipmunk is a blind, hairless
gobbet weighing perhaps a tenth of an ounce; but in a
week its striped coat begins to grow,
in three more its eyes open, and after five or six it
is weaned. The adult in the large portrait
is setting out to forage, while the other
one is cramming a morsel into an already bulging
pouch. Its home tunnel may be thirty feet
long and invariably contains storage chambers as well
as piled seeds and other gleanings in a
nest chamber. The animal will sleep through cold
winter weather, but during warm spells it will partake
of the stored food and occasionally come out to
poke about or bask briefly in the sun.

PLATE

39

Thirteen-lined Ground Squirrel

Spermophilus tridecemlineatus

Sundry members of this genus, scattered across western America, were known to Audubon and his contemporaries as spermophiles, "seed-loving creatures." Mammalogists later grouped them with Old World relatives in the genus *Citellus*, then separated them again and reverted to the designation *Spermophilus.* Early descriptive ambiguities led to a succession of four Latin names for one Mexican species. Another, the ringtailed ground squirrel, was described by Audubon and Bachman in 1842 from a single sample donated by an adolescent amateur naturalist, Spencer F. Baird, who had acquired it from a dealer. Years later, when Baird headed the Smithsonian Institution, he concluded that his ringtailed squirrel must be an unfamiliar African rodent. Its origin remained doubtful until a second specimen was collected in Mexico in 1877. Much more widely distributed is the thirteen-lined variety, common from the Rockies to Ohio. Naturalists of Audubon's time called it the "leopard spermophile" in metaphorical tribute to "its gaudy dress." The animal's habitat transforms gaudiness into camouflage as its pale and spotted dark stripes blend into the grasslands.

No. 8.

SPERMOPHILUS TRIDECEMLINEATUS, MITCHELL.
LEOPARD SPERMOPHILE.
Natural Size
MALE AND FEMALE.

Thirteen-lined ground squirrels generally have six- or seven-inch bodies and four- or five-inch tails, but local groups and geographic races may differ somewhat in the tone of their striped and dotted coats since they must blend with the environment. The tawnier of the two presented here is an Ohio specimen, while the grayer one is Kentuckian. An extremely intense hibernator, the thirteen-lined species plugs its burrow in winter, rolls up, stiffens, and reduces its allegro respiration to one breath in five minutes. The shorter-tailed, heavier-bodied, solidly colored Richardson's ground squirrels *(S. richardsonii)* in the top photograph inhabit grain-rich western and midwestern plains from Canada to Colorado. Sitting up and barking like a prairie dog is an arctic ground squirrel *(S. undulatus),* found from Alaska to Hudson Bay. Largest of the genus, it escapes danger by plunging into a burrow as its smaller relatives do, but when cornered it can sometimes stave off an enemy as large as a fox.

No 20.
SPERMOPHILUS LUDOVICIANUS, ORD.
PRAIRIE DOG - PRAIRIE MARMOT SQUIRREL.
Natural Size

PLATE XCIX.

PLATE

99

Black-tailed Prairie Dog

Cynomys ludovicianus

Canadian voyageurs impressed by the barking of large, rotund, tawny, burrowing squirrels named the animals *petits chiens,* and the vast mounded colonies were commonly called dog towns by the time Audubon investigated some of them near the Missouri in 1843. He reminded readers that Lewis and Clark had poured five barrels of water into a typically extensive den without filling it, and abandoned an attempt to excavate one upon discovering that the plunge hole descended more than twelve feet. As recently as 1901 a Texas dog town pocked an area of twenty-five thousand square miles, but grazing cattle now compete with few prairie dogs where once there were hundreds of millions. Audubon depicted one of three black-tailed subspecies found on the plains from Canada to Mexico. There are also four races of white-tailed prairie dog, C. *leucurus,* in the high meadows of the Rocky Mountain region from Montana southward. Those of the flat plains are the mound builders, diking their shafts against flooding with cratered domes that double as sentry platforms. Prairie dogs stand erect on some of the domes, which are often a yard high and ten feet across. When an invader such as a hawk, coyote, or badger elicits yelps of alarm, feeding and sunning prairie dogs race for their holes and dive out of sight.

Lith. Printed & Col.d by J.T. Bowen Philad.a 1846.

On a burrow's dome gather half
a dozen prairie dogs—most likely a female
and her adolescent young. They
will receive many visits, particularly
from the coterie's dominant male.
A dog town tends to disperse
into villages, bounded by features of
terrain and vegetation, and
these sections are subdivided into
coteries that typically include
several females, one or two males, and the
young produced in the last two
years. Gregarious and affectionate
behavior notwithstanding,
border disputes must be quelled by a
dominant male. When not
yelping, visiting, sleeping underground,
or watching for predators, the
residents spend most of their time
eating preferred grasses
and other plants, or cutting down the
less savory stalks to ensure
a clear view of approaching enemies.

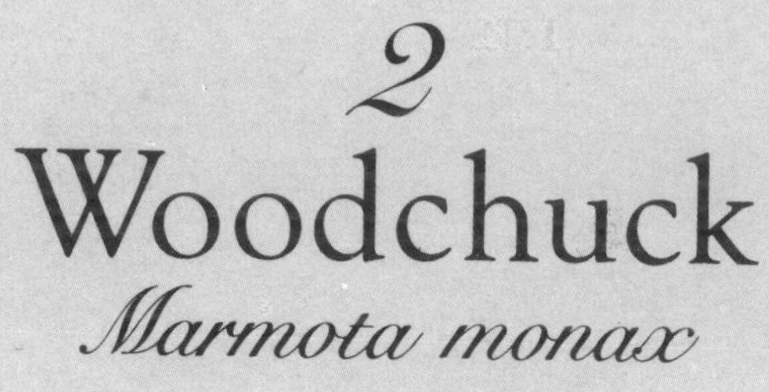

PLATE

2 Woodchuck

Marmota monax

Wuchak—"woodchuck"—is an Indian word that holds no clue to habitat. The Crees evidently used the name casually to identify the marmot and other mammals having little in common but general size and color. This species usually spurns woods for woodland edges, clearings, fields, and pastures. The name groundhog better fits this squat, portly animal that waddles from its burrow to munch insatiably at grasses, herbs, vegetables, and fruits. The nine subspecies are most thickly

Drawn from Nature by J.J. Audubon, F.R.S F.L.S

ARCTOMY.

MARYLAND, MARMOT

PLATE, II.

distributed over the eastern United States, but reach west to Idaho, across all of Canada, and into Alaska. The supposedly eastern woodchuck is sometimes mistaken for one of the western rockchucks, the yellow-bellied marmot *(M. flaviventris)*. All of the species hibernate. A rabbit may take shelter in a woodchuck tunnel, but the nest chamber is plugged in November and its occupant sleeps undisturbed until early spring. Thus woodchucks escape cold and hunger, as Bachman observed, for they "have no winter in their year."

Lith. Printed & Col.d by J. T. Bowen, Phila. 1842.

AX GMEL.

ICK, GROUNDHOG.

Though a woodchuck appears complacent when it sits up, the animal is scanning its environment for signs of danger. Its hearing is acute and its eyes can detect movement seven hundred yards away. The one at right, a female, may well have a quartet of infants in her burrow. They are usually born in April or May. Before June has passed, the mother leads each one away to its own first den, which she has prepared in advance. By midsummer a typical juvenile migrates to a new field where it digs a burrow for itself. It will grow fat on many kinds of pasturage. The trencherman at left is chewing a plantain leaf. Below, a woodchuck emerging from its hole shows the prominent incisors that characterize rodents. Since these teeth never stop growing they must be ground down constantly, and malocclusion is an occasional cause of death.

No. 21.
Drawn from Nature by J. W. Audubon.
ARCTOMYS PRUINOSUS, PENNANT.
HOARY MARMOT - THE WHISTLER.

PLATE CIII.

Lith. Printed & Col. by J.T. Bowen, Philad., 1846

PLATE

103
Hoary Marmot
Marmota caligata

Among North American rodents only the porcupine and beaver outweigh this species, the largest of the marmot tribe. A hoary marmot occasionally grows to a length of more than two-and-a-half feet and a weight of nearly twenty pounds. Mammals in the North tend to be larger than their southern counterparts; this one inhabits alpine meadows and rocky slopes from Montana, Idaho, and Washington up into Alaska and the Yukon. Though coloration varies among individuals as well as among the ten subspecies, a hoary marmot always has black feet—giving it the booted look denoted by its Latin name. Its coat is usually grayish, whitely flecked and frosted, with more pronounced black-and-white mottling about the head and shoulders. All marmots utter a short whistle of alarm, and the largest is also the loudest. Ever since Audubon's time, it has been known as the whistler, or *le siffleur.* Wind may carry the shrill call for a mile, alerting voles and wild sheep, lemmings and deer, and other prey large and small to the patrol of some unseen carnivore.

At ease on a high Alaskan ledge or shadowed
by the peaks of the Rockies, a hoary marmot shares the
domain of wild sheep and mountain goats.
Like those animals, it looks downslope for ascending
enemies or faces into the shifting wind to
catch alien scents and sounds. This is why three or four
sunning marmots, using a jumble of rocks
as an observation platform, will all tend to gaze in the
same direction. Both this species and the
yellow-bellied marmot are commonly called rockchucks.
Hoary marmots, especially, favor talus slopes
where little mountain pastures offer grasses and forbs,
never more than a short dash from the safety
of rocky clefts. Sometimes they pock the meadows
with burrows, in the manner of woodchucks, but they also
use natural rock dens that are more secure
from the bears that hungrily dig for them in the spring.

PLATE

123

Mountain Beaver

Aplodontia rufa

Uniquely American and the sole member of its family, the mountain beaver cuts and cures its fodder and digs labyrinthine tunnels only in humid forests along the Pacific Coast, from southern British Columbia to northern California. It is not a beaver, nor is it a marmot though it looks vaguely like a small, blunt-headed, tailless woodchuck. The species is the oldest living rodent, and it has been a resident of this continent for at least sixty million years. It may be found among mountains, but rarely on high slopes, for it requires a mild climate, heavy vegetation, and plenty of water. When its tunnels flood, it splashes contentedly through the underground rivulets to and from its nest chamber. Bachman acknowledged Lewis and Clark as the discoverers of this rotund curiosity. It measures a foot or a foot and a half in length, weighs from two to four pounds, and has a pale spot under each ear and a very stubby tail hidden in coarse fur. Adopting a Chinook term for robes that often utilized the pelts of this species, Lewis and Clark called their discovery the sewellel. This name and several others remain current, but some naturalists avoid confusion by using only the name aplodontia.

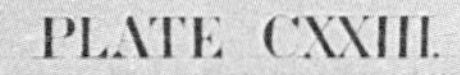

APLODONTIA LEPORINA, RICH.

THE SEWELLEL.

MALE.

Natural Size.

Lith^d Printed & Col^d by J.T. Bowen Philad^a 1847

Visible through the dry grass at left is one of the kangaroo rat's very long hind legs, its only real defense. The animal is seen in a characteristic stance, with its tiny forelegs tucked up against its chest, ready to spring and dodge on its hind legs at any hint of danger. Similarly poised is the one pictured above. This is the little Merriam species studied by John Woodhouse Audubon. It stands on a mat of litter that may mark a tunnel entrance. Unlike most kangaroo rats, however, this kind stores no food in its burrows and only a little on the surface. It would rather court a beating by pilfering silage from larger relatives. The tasseled tail of the kangaroo rat at right seems to be sweeping the desert sands as it swishes restlessly near a burrow where three or four young may wait to be nursed.

PLATE

36

Porcupine

Erethizon dorsatum

Hidden amid underfur and long, coarse guard hairs from the top of a porcupine's blunt head to the end of its thick tail are about thirty thousand barbed quills, some of them three inches long, an armor "bristling with bayonets," as Bachman described it. Normally sluggish, a threatened porcupine growls, clicks its teeth, tucks its vulnerable face between its forelegs, turns its rump toward the foe, arches its back, and whips its tail from side to side. The loosely anchored quills easily pierce skin and muscle, the barbs penetrating more deeply as the victim tries to rub them away. Enemies as large as bears may starve when a severe harpooning in and about the mouth prevents feeding. Only a few carnivores, notably the fisher, are adept at flipping a porcupine over to attack the unshielded belly. The species is found throughout most of Canada and the United States, except for the South and parts of the Midwest. When a porcupine is mature, in its third year, it usually weighs about twelve pounds, but may grow to three times that weight and a length of three feet or more. Occasionally seen during daylight on the ground or perched on a high branch, the animal prefers to feed at night, chiefly on the cambium of both hardwoods and conifers. It can kill trees by girdling them, and may further incite the wrath of farmers and woodsmen by gnawing through axe handles, canoe paddles, and other tools to satisfy a craving for salt.

Nº 8.
PLATE XXXVI.

A female porcupine bears her single
infant in late spring, after a seven-month
gestation period. At birth it is open-eyed, haired,
and armed with quills that harden as
they dry. It may be born in a tree cavity or a
hollow under rocks, and as an adult
it will shelter in a den during severe weather, but
it remains active all winter, subsisting
chiefly on bark. The one pictured on a seemingly
dead, snow-flecked limb may be resting
or foraging. Another is seen in summer, feasting
on large-toothed aspen leaves. Porcupines
are slow but sure-footed aerialists, unafraid to
doze on precariously thin branches.
They are also competent swimmers, buoyed up by
their stiff, hollow quills. The one
on the gravelly sandbar is in its normal walking
position; if alarmed, it would flatten
itself to keep its vulnerable belly to the ground.

PLATE

46 Beaver

Castor canadensis

A marvel of adaptation, the largest American rodent has oversize lungs that enable it to swim half a mile underwater. Its massive hind feet are webbed, its ears and nostrils valved to close on submerging, its lustrous chestnut fur waterproofed by oil. Its big orange incisors can fell a young cottonwood, a poplar, or a fully grown willow in minutes. An adult beaver, weighing fifty or sixty pounds, uses its broad, flat, hairless, scaly, powerful tail, frequently over a foot long, as a rudder, for balance and

PLATE XLVI.

support when reaching up, and to slap the water,
sounding an explosive alarm. Once devastated by the fur
trade, beavers have been restored to most of their
transcontinental range. They eat aquatic plants, grasses,
and the bark and twigs of the trees they cut. With
branches, saplings, and mud they dam streams
to form ponds in which they build domed lodges—moated
citadels with underwater tunnels ascending to
a dry chamber where a monogamous pair nests with three or
four infants and often a like number of yearlings.

On a grassy bank, a beaver
sits up with its flat tail tucked
forward between its legs
while it devours a twig. The one
in the water is stripping
leaves from a submerged aspen it
has felled. Another, about
to reach for twigs on
a standing sapling, balances on
a tail three inches wide.
The tails of large beavers may be
broader by half.
Above, a lodge is being repaired.
Bachman observed that the
male leaves the nest—which may
be such a lodge or a burrow
in the bank—before the kits are
born, and returns in
midsummer, when they are weaned.

PLATE

13

Muskrat

Ondatra zibethicus

Algonquian Indians called the animal *musquash*, a word designating the reddish tinge often seen in its brown fur. Perhaps inevitably, the name evolved into "muskrat," for the species is a rodent that exudes a mate-attracting musk. Though it gnaws down no trees, it looks and acts vaguely like a miniature beaver. Its head, however, is less blocky, its hind feet only partially webbed, and its tail—flattened vertically rather than horizontally—is pointed and not nearly so broad. An adult is about a foot long, with a tail measuring almost another foot. Thriving in marshes, rivers, ponds, and lakes from the Arctic to the Rio Grande, it is absent from only small portions of the United States. Whereas beavers generally dig burrows in a bank only where the water is swift, deep, or unstable, muskrats almost universally burrow, and they also build water houses—smaller than beaver lodges—of reeds and grasses. They will eat almost any available vegetation and supplement this diet with small fish, crustaceans, and other minuscule prey. They are voracious feeders that, in turn, provide forage for mink, raccoon, and similar predators capable of raiding nests in search of the young.

PLATE XIII.
FIBER ZIBETHICUS, CUVIER.
MUSK-RAT, MUSQUASH.
OLD AND YOUNG.
Lith. Printed & Col.d by J. T. Bowen, Phila. 1843.

The half-dozen muskrats in a typical litter can swim before their eyes open. The *Quadrupeds* called water "their proper element" and described them "on a calm night in some mill-pond or deep sequestered pool, crossing and recrossing . . . leaving long ripples . . . whilst others stand for a few moments on little knolls . . . or on stones or logs." They prefer spring-fed, marsh-lined waters that seldom freeze over, but they will graze in a frigid Alaskan pool or rest on the slabs of melting ice that clutter a New Jersey shore.

No. 24.

Drawn from Nature by J. J. Audubon F.R.S. F.L.S.

Fig. 1. GEORYCHUS HELVOLUS, RICH. TAWNY LEMMING. Natural Size

Fig 2. & 3. GEORYCHUS TRIM BACK'S LEMMIN Natural Size

PLATE

120
Brown Lemming
Lemmus trimucronatus

The stubby-tailed yet mouselike lemmings exist in a frenzy of foraging, fleeing, and reproducing as they provide food for hosts of fur bearers and predacious birds. Audubon and his younger son painted all three American groups: the brown, or common, lemmings of Alaska and upper Canada; the collared, or varying, species of the more remote Arctic; and the bog lemmings (not true lemmings anatomically) of Alaska, Canada, and the eastern United States. All have short legs, big heads, and thick, coarse fur that hides their small ears. The brown lemming, sometimes seven inches long, is the largest. Cutting runways through tundra and taiga or furrowing through snow, lemmings eat seeds, grass, berries, roots, mosses, and lichens. Their northern populations peak every three or four years. By the next summer they demolish their browse and dash about aimlessly, though without staging the vast migrations of the closely related Norway lemmings. The hordes are ravaged by starvation, disease, and feasting predators. As a new cycle begins, bird watchers in the United States are startled by an unusual influx of snowy owls, seeking food to replace a staple in short supply.

A mat of lichens, mosses, and sedges is ideal habitat for the lemming. However, the Labrador collared, or varying, lemming *(Dicrostonyx hudsonius)* prefers less boggy tundra than the brown lemming. The collared species is shown below nibbling lichen from a boulder. Marked by a slight ruff behind its neck, it is the only rodent that turns white in winter. Both this species and the brown lemming use surface nests as well as burrows. The ball of grass at left, beside a tunnel entrance, is a moss-lined nest about six inches in diameter. In the bottom picture, a snowy owl offers its mate a brown lemming. For these raptors and others, lemmings provide an indispensable food source.

PLATE

40

White-footed Mouse

Peromyscus leucopus

Audubon and Bachman correctly believed the white-footed mouse to be the most widespread of all native American mice, though they failed to perceive that specimens donated from many distant points represented a genus rather than a single species. In addition to the prototype *P. leucopus* of Audubon's plate, there are the deer mouse, cotton mouse, canyon mouse—twenty-seven species in all, a hundred and fifty subspecies—differing slightly in color and anatomy, and adapted to diverse habitats throughout North America. The species depicted dwells in woods and fields across the eastern two-thirds of the continent. Its drably grayish young develop into elegantly fawn-red or buff adults with white feet and underparts. The tail is about as long as the head and body, three inches or so, and often as white beneath as the belly. This mouse subsists chiefly on seeds, nuts, and insects. It grooms itself meticulously but is so untidy that it must frequently build a new nest of shredded vegetation and miscellany in a hole or cranny, under a tree, among rocks, even in a cabin. A good climber, it often roofs over an abandoned bird's nest, "with as much art and ingenuity," Bachman declared, "as the nests of the Baltimore Oriole."

PLATE XL.
MUS LEUCOPUS, RAFF.
WHITE FOOTED MOUSE.
MALE, FEMALE AND YOUNG.
Lith Printed & Col. by J. T. Bowen, Philad. 1844.

In an inconspicuous
hole beneath forest litter, a
white-footed mouse
nurses her gray-coated young.
They are about three
weeks old and ready to be weaned.
As with most small mammals,
infant mortality is
high. These are survivors of a
litter that probably
numbered five. A plump one-ounce
adult spends much time
investigating its surroundings, as
in the bottom picture,
and grooming itself, as at
right. Pictured below is another
member of the genus, the
deer mouse *(P. maniculatus),* which
shares the woods and
prairies with its white-footed
relative. Either species
occasionally settles in a cabin
or quiet rural house.

No. 9.
PLATE XLV.
Arvicola Pennsylvanicus, Ord.
Wilsons Meadow Mouse
Natural Size

PLATE

45

Meadow Vole

Microtus pennsylvanicus

Meadow voles, known in many regions as meadow mice, probably are the continent's most prolific mammals. In a temperate climate a female that begins breeding when twenty-five days old may continue to do so through the winter, easily producing fifteen or more litters—perhaps a hundred young—in a year if she lives that long.

Like lemmings, white-footed mice, and other small rodents, voles are hunted by virtually every carnivore. After depicting the meadow vole, Audubon also painted the pine vole, yellow-cheeked vole, and water vole. Scores of other species blanket the continent, some smaller, some slightly larger, some nearly identical but for skull structure and coloring. The meadow vole is a plump little animal about seven inches long—including two inches of slightly furred tail. Its head is large, its snout blunt, its ears almost hidden in fluffy, grizzled fur, grayer in the West, browner in the East. It is the most widespread of voles, common from the upper two-thirds of the United States to the Arctic. An eater of grass, bark, roots, and seeds, it tends to burrow in summer, but builds winter nests in weed clumps or under debris. It cuts inch-wide trails in the grass, often so industriously as to form a maze. In its quest for food it tunnels through snow and occasionally slices out a hatch to surface, never knowing if a predator awaits it. Whereas a white-footed mouse drums with its front feet when alarmed, a vole stamps its hind feet in the manner of a rabbit. Evidently this warns its neighbors to stay hidden, for the species suffers such constant predation that survival requires the escape and reproduction of immense numbers.

Meadow voles are active
day and night. On an autumn day, a
quiet observer may see one
pop out of its hole, perhaps in
a knob of earth or under
roots. It will hustle along, mowing
a tiny path by cutting plant
stems off at the ground.
The soft or tasty parts are eaten,
the rest stacked by
the trail. The animal's daily
consumption of food may
total several ounces—nearly its own
weight. As winter approaches,
it will probably vacate its burrow
and weave a grass nest in
a clump of weeds or beneath debris.
Then it will resume
foraging, often under snow
that provides concealment as well as
a roof of insulation.

PLATE

4

Eastern Wood Rat

Neotoma floridana

Although similar in shape and size to the Norway rat regrettably introduced by the Colonists, the seven species and many subspecies of native wood rats are handsome and harmless. Their soft fur ranges from buffy brown to gray, sometimes sprinkled with cinnamon, and their underparts and feet are white, their tails of moderate length and haired. From Canada to Mexico there are varieties adapted to deserts, forests, plains, and mountains. The eastern wood rat is distributed from the Dakotas to the Atlantic and southward to the Gulf. Audubon also depicted the bushy-tailed wood rat *(N. cinerea),* a western form instantly recognized by its squirrel-like thatch of tail. All wood rats construct large ramshackle nests—sometimes three feet across—of sticks, cactus needles, cane, or whatever materials are available. The eastern type often builds its home in a rock crevice or a tree, and is likely to have a separate granary for the storage of seeds, nuts, berries, and grass. In addition to these staples, insects are consumed. The bushy-tailed species is the pack rat, or trade rat, of miners' stories about pilfered spectacles, jewelry, false teeth, and dynamite sticks; but the eastern wood rat and all other species share the same odd habit of stealing away in the night with a bright or oddly shaped object, often leaving a pebble or twig in its place. They habitually carry debris to reinforce the nest, and "trading" occurs each time they drop what is in their mouths to pick up a more attractive object.

No. I.
PLATE IV.
Drawn from Nature by J. J. Audubon, F.R.S. F.L.S.
Lith. Printed & Col.d by J. T. Bowen, Phila.
NEOTOMA FLORIDANA. SAY ET ORD.
FLORIDA RAT
Natural Size
MALE, FEMALE AND YOUNG OF DIFFERENT AGE

An eastern wood rat
is pictured sifting through a mat of
brown and scarlet autumn
leaves and spore-laden ferns. It may be
uncovering edible seeds,
but is probably looking for nest material to
be carried into a nearby recess
in the rocks. At warmer times of year a female
might be preparing a nest
for the birth of her two or three young.
They will spend most of their
first few weeks nursing, and if the mother
senses danger she will drag them,
clinging to her nipples, to a safer hiding place.
The other animals pictured represent two
races of the closely related
western bushy-tailed species popularly known as
the pack rat. These clean, graceful
creatures have copiously furred tails and long,
sensitive whiskers that jut out and
up as jauntily as a waxed mustache. The one at
left is on a new nest, to which will
be added a protective entanglement of sticks—
virtually an abatis—and perhaps
a decorative welter of bright objects.

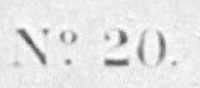

PLATE

98 Ringtail

Bassariscus astutus

The ringtail, or cacomistle, is a shy carnivore with the characteristics of more familiar species and a host of allusive names: squirrel cat, ringtailed cat, civet cat, raccoon cat, and raccoon fox, among others. Related to the raccoon but smaller and slimmer, it weighs two or three pounds and stands six inches high at the shoulder. Its big ears nearly frame a delicate, foxlike face. The brownish-gray coat pales to whitish spectacles about the eyes, and the long, luxuriant tail has black and white bands, usually fourteen, with the black fading away on the underside. The large eyes are adapted to nocturnal hunts for rodents, birds, lizards, frogs, and insects. An opportunist like the raccoon, a ringtail will also eat fruits, nuts, corn—almost anything edible. Three subspecies range from Mexico through the American Southwest to California and lower Oregon. They den in cliffs, between or under rocks, and in tree holes. Among the live oaks and post oaks of Texas, John Woodhouse Audubon observed their habit of gnawing the bark around their dens, a welcome aid to a naturalist seeking the species. Usually three or four young, white-haired at first, are born in spring or early summer. The sire stays away for three weeks but then, like a fox, joins his mate in bringing food for the litter. After four months a young ringtail can cope with a solitary existence. As John Audubon remarked, it leaps well and climbs with the "agility and grace" of a squirrel.

One of the ringtail's many names is
cacomistle, a derivative of an Aztec term meaning
half-cougar. An equally popular name is ring-tailed cat,
and indeed the Utah specimen at left and in
the bottom picture does appear feline
in the way it climbs, pauses, and crouches while prowling
about a log in quest of food. Unlike a
cat, however, it will happily eat fruits or nuts,
particularly acorns, if it finds no rodents or reptiles,
no birds or bats or other prey. Another
ringtail, suspending a search for insects in a sprinkle
of forest litter on a boulder, appears to be
barking. The species is not vociferous, but its young
can squeak and an adult will snarl or bark
when excited. A ringtail may also scream when fighting
or fleeing an enemy such as an owl.

PLATE

61

Raccoon

Procyon lotor

N°13.

Though scarce in the Rockies and unknown in desert regions, nineteen raccoon subspecies flourish from Canada to Panama. Audubon evokes a northern raccoon in autumn: corpulent but alert, perhaps weighing close to twenty pounds, its ruff already winter-thick, body and ringed tail bushily coated, masked face a blend of shrewdness and innocence as it bellies along a lichen-dappled log in quest of food. Bachman and Audubon recorded a few items in its diet—birds' eggs and nestlings, frogs, turtle eggs, mussels, ducks, corn—a list barely implying the animal's omnivorous appetite when it prepares to den and doze through winter's coldest weeks.

During the autumn feeding frenzy
a large male raccoon has arrived in a field to
steal corn but is distracted
by good fortune—a covey of quail. A cock
bobwhite has become aware almost
too late that it is being stalked, and now it
runs for its life. Raccoons
tend to be solitary unless a rich food store
draws them together, but in late winter
or early spring a mating pair will
share a den tree for a week or so. Four or
five young are born two months later.
Above, a juvenile has become soaked during an
excited attempt to pin a crayfish,
frog, or other small prey.

PLATE

141
Black Bear
Ursus americanus

Audubon and Bachman astutely declared the black bear and the "cinnamon bear" to be one species. The lustrous black prototype claiming a deer leg in Plate 141 usually has a tan snout and often a pure white chest blaze. Even in the East, where this color form prevails, twins or triplets may differ. Cinnamon, tan, brown, or blond becomes prevalent to the West. More startling hues mark two of the eighteen subspecies, the whitish Kermode race *(U. a. kermodei)* of west-central British Columbia and the blue-gray or silvery blue-black glacier bear *(U. a. emmonsii)* of Alaska's gulf coast. The *Quadrupeds* devoted a second plate to the cinnamon variety, "not because we felt disposed to elevate it into a species, but because it is . . . so frequently found in the collections of skins . . . and . . . so often noticed by travellers in the northwest, that errors might be made by future naturalists were we to omit mentioning it." Equally knowledgeable was the description of the rarely seen newborn young as "exceedingly small." At birth a cub seldom weighs much over eight ounces, though three years later it may weigh three hundred pounds and a few old males weigh twice that. Probably the world's most abundant bear, it is a purely American species, a shambler through forests from Alaska and Canada to northern Mexico.

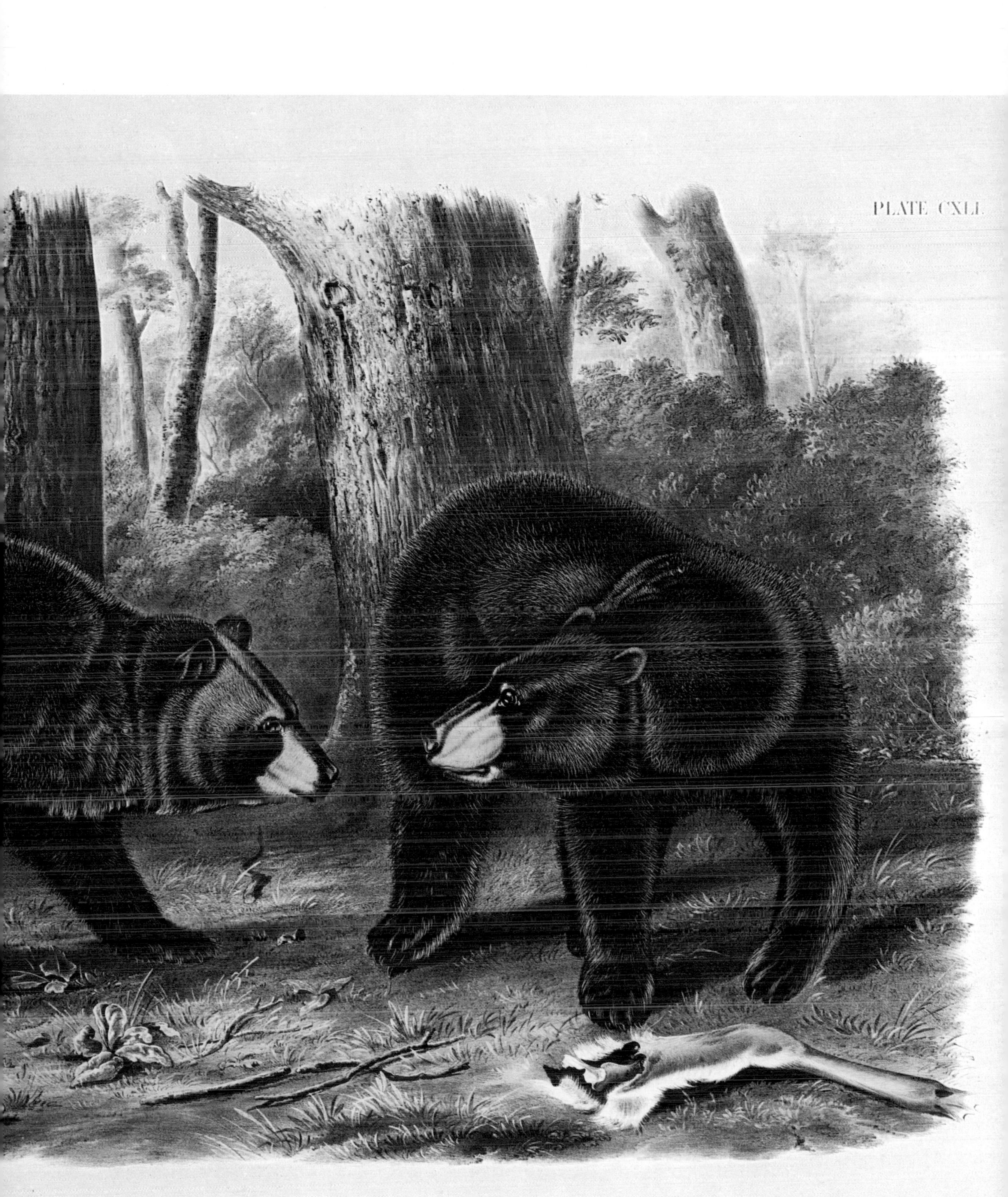
PLATE CXLI.
URSUS AMERICANUS, PALLAS.
AMERICAN BLACK BEAR.
Lith. Printed & Col. by J. T. Bowen Philad. 1848

The continent's only ursine species that retains its tree-climbing ability after maturing, the clumsy-looking black bear is remarkably agile for its large size. It can turn on an adversary with blurred speed, as at right. It eats far more vegetation than prey but will battle smaller rivals for a kill or a bit of carrion. The glistening black coat and tannish snout patches of the bear in the top portrait are typical in the East, yet various shades of brown are more common in other regions. And the cream-white specimen above is a normal black bear of the Kermode race found in west-central British Columbia.

wn from Nature by J.W. Audubon.

URSUS-FEROX, LEWIS & CLARK.

GRIZZLY BEAR.

MALES.

PLATE

131
Grizzly Bear
Ursus arctos

All beasts defer to an approaching grizzly. Audubon professed difficulty in describing "sensations experienced on a sudden . . . face-to-face meeting with the savage . . . shaggy monster disputing possession of the wilderness." Yet the plate shows two grizzlies in gentle summer companionship, for a mating pair may dally amorously for weeks though at other times they are apt to be hostile even to their own kind. As with black and polar bears, delayed gestation brings forth the young while the sow dozes fitfully in a winter den. The grizzly has European and Asian counterparts but human encroachment has spared only two American races, the interior grizzly (*U. a. horribilis*) and Alaskan brown bear (*U. a. middendorffi*). Many interior grizzlies are silvertips and many are blond or coppery or as brown as the coastal Alaskan type. A few remain in Wyoming and Montana, far more in Alaska and western Canada. On a diet of grasses, berries, mast, tubers, insects, carrion, and prey, a grizzly may grow to seven hundred pounds or more. A coastal brown, fattened on spawning salmon, may reach twice that weight, vying with the polar bear as the world's largest carnivorous mammal.

Its teeth lightly flecked
with blood, a feeding grizzly raises
its massive head to test for
the scent of an intruder. The skull of
an interior grizzly like this
one may be nearly eighteen inches long
and ten inches wide; that of a
coastal Alaskan brown may be slightly
larger. The browns at left,
fishing for spawning salmon, look
peaceable, but a vicious fight may ensue
if one of them crowds the dominant
bear facing upstream. The
grizzly sow above is resting on a bank
while her twin cubs nurse.

PLATE

91
Polar Bear
Thalarctos maritimus

Craning its long neck over the brink of a precipice, Audubon's polar bear scans an Arctic sea for food —a seal on a floe, an injured bird, a dead fish, or nondescript carrion. The great white bear can dive from a fifty-foot cliff, though more often it hunts hair seals by stalking the edges of ice fields or lying in wait at plunge holes. Circumpolar in distribution, the species is found chiefly on sea ice, coasts, and islands, but pregnant sows often venture inland to den. In 1833, while cruising Labrador's coast, Audubon searched unsuccessfully for the bears. They regularly wander that far south, and still farther down the coasts of James Bay. Many males, however, spend their adult lives on sea ice, seldom or never denning and perhaps never seeing land. They swim longer distances than any other large quadrupeds and have been sighted two hundred miles out in the ocean. Thick tallow and extremely dense fur make them impervious to cold. Cubs trailing a sow or clinging to her rump as she swims are pure white except for claws and pads, eyes, nose, and lips. Eventually their fur acquires a yellow cast, and some old bears weigh over half a ton, yet they can be almost invisible slumbering amid ice heaves and snow dunes.

PLATE XCI.

URSUS MARITIMUS, LINN.

POLAR BEAR.

MALE.

Lith.d Printed & Col.d by J. T. Bowen, Philad.a 1846.

As a sow polar bear and her two-year-old
twins lope across wind-sculptured ice on Alaska's
northernmost point, an observer might
have difficulty telling the cubs from their mother.
A female seldom weighs much more than seven hundred pounds.
This one may be less than six years
old and still growing if the cubs are her first young.
They have learned to fend for themselves
by now, and she will soon abandom them and breed
again. Often, a polar bear shakes
itself like a dog as it leaves the water. Little moisture
can penetrate the thick fur, but an encumbering
cloak of icicles might form if the outer coat remained
very wet. In summer the bears wander
slightly inland to graze on sedges and grasses, hunt for
injured prey, feast on carrion, and
occasionally devour the chicks of ground-nesting birds.

No. 11
LUTRA CANADENSIS, SABINE.
CANADA OTTER.
MALE.

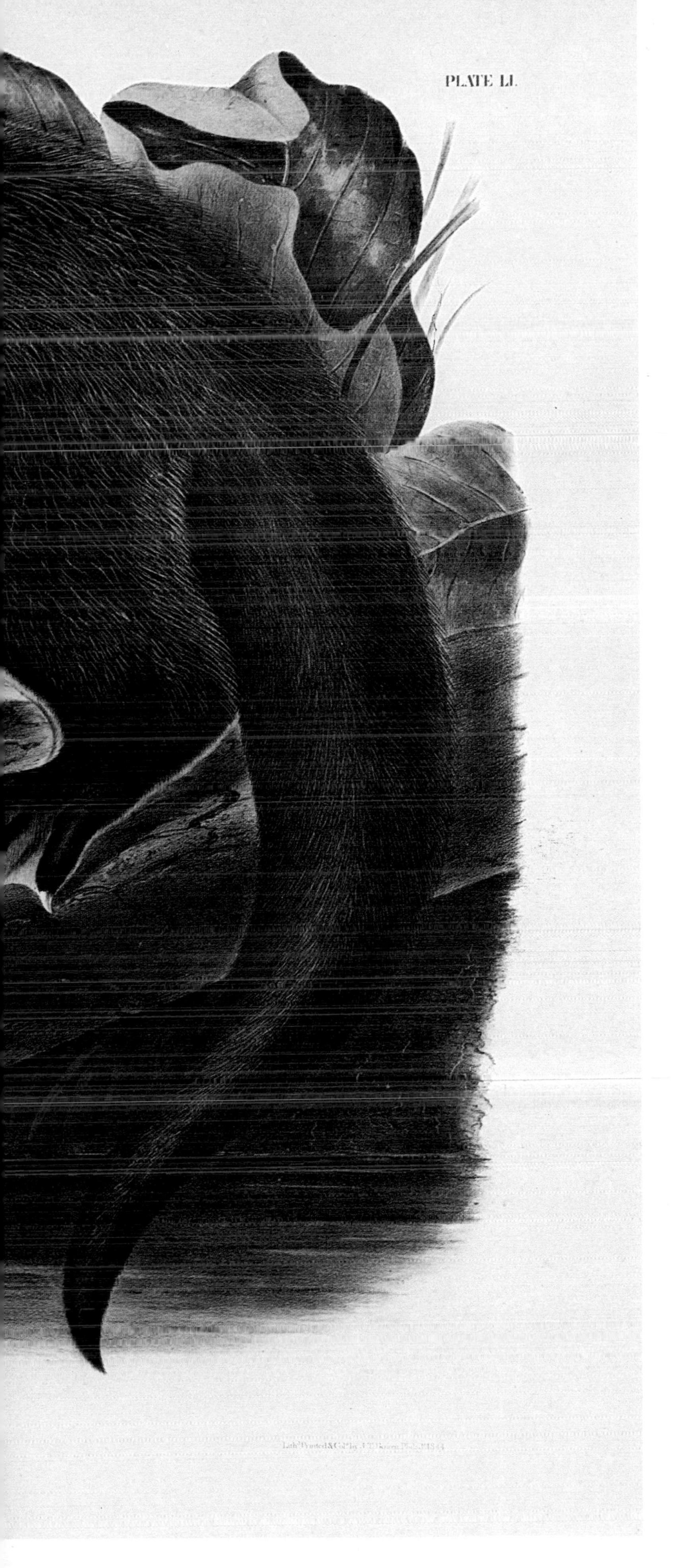

PLATE

51

River Otter

Lutra canadensis

North to the Arctic and down through all of the United States except arid portions of the Southwest, the river otter plies streams and ponds for crustaceans, mollusks, fish, and other small prey. Its numbers on this continent have been reduced by pollution, river traffic, and, until recent years, by extensive trapping. Its durable pelt, richly dark and glossy, was even more valuable than the beaver's. Anglers in some regions condemn the remaining otters, but the species kills fewer trout and bass than is commonly supposed, as it prefers slower prey. Audubon painted this otter from sketches made forty years earlier, when he shot a specimen on the banks of the Ohio. He called the portrayal an endeavor "to represent the pain and terror" of a trapped otter. A large aquatic weasel sometimes weighing over twenty pounds, the otter is the only web-footed member of its family. Its thick, tapered tail serves as a scull and rudder, and its ears and nostrils are valved like the beaver's. It is among the most playful of all mammals, frolicking in the water with its two or three pups or its mate and tobogganing down mud slides or snow chutes formed in steep banks.

In an Everglades pool, a river otter holds a fish that may have been caught broadside but was flipped around for lengthwise swallowing. Other prey includes worms, salamanders, frogs, crayfish, snails, clams, snakes, turtles, insects, and now and then a small mammal or bird. An otter may cruise on the surface, as at right, or submerge for several minutes. Chirping, grunting, and snorting on a bank like the one above, otters make mudslides down which they slither and skitter.

PLATE

137 Sea Otter

Enhydra lutris

Before the carnage was halted by international treaty in 1911, hundreds of thousands of sea otters were killed for their pelts—velvety, thick, silver-glazed, varying from mahogany to blackish-brown, often paling to gray or gold about the head. It is the world's most lustrous, durable, valuable fur. Pods of otters no longer bob in the shallows along many Pacific coasts, although the slaughter stopped short of extinction. Protected colonies exist off lower Alaska, the Aleutians, California, and in Siberian and Japanese waters. Sea otters attain a length of about four feet, and some of the northern males grow even larger on a diet of mollusks, sea urchins, crabs, and fish. They are superb divers, equipped with flipperlike hind legs and webbed toes. They breed in the water, beaching in foul weather but eating and sleeping afloat on their backs.

To crack a mussel or abalone, an otter places a stone on its chest as an anvil. To anchor itself before sleeping, the animal wraps a kelp strand around its body.

Bachman and Audubon doubted a report that wary otters habitually raise their heads from the water and shade their eyes with their forepaws to scan the surroundings—doubt was reasonable but the report was true.

A sea otter floats
on its back and holds its
food between its
paws. This is also the
position in which
a female nurses her single
young, born in the
spring. While eating soft
prey such as squid,
an otter sometimes clasps
a rock between
its body and one foreleg,
thus reserving a
mollusk-smashing tool for
future use. Even a
very large male, perhaps
weighing close to
eighty pounds, can float
almost indefinitely.
The otters pictured above
on an Alaskan beach
are waiting out a gale.
When it subsides, they will
return to the water.

No. 28.
Drawn from Nature by J.W. Audubon
MUSTELA MARTES, LINN.
PINE MARTEN.

PLATE

138
Pine Marten
Martes americana

The pine marten, like most other members of the weasel family, is slender, long-bodied, short-legged, and equipped with anal musk glands. The fur varies from nearly blond to dark brown—it pales on head and belly and darkens on tail and legs. An adult is usually about two feet long, the tail accounting for a third of that length. The species looks rather like a mink or fisher, though the fisher is bigger and more uniformly dark, and the mink darker, shorter-tailed, and white-throated. Related to the Russian sable and often called American sable by furriers, it has suffered heavy losses through trapping as well as disruption of habitat by lumbermen, but it is now protected in many areas. It dwells in the evergreen forests of Alaska, Canada, the upper Appalachians, the Rockies, and the Pacific coastal mountains as far south as central California. Darting through treetops or ground cover, it swiftly and relentlessly hunts squirrels, mice, birds, and eggs, enriching its diet with berries, cone seeds, and honey. It might easily catch a junco, as in the plate, but is a solitary prowler that does not share its kill and rarely has to guard it from another marten.

Surprised in her nest, a female Colorado marten snarls and bares her teeth in a threat display. It is early spring and she is preparing this nest in a hollow tree for the birth of her litter, due after a gestation period almost equal to that of the human species. As with other weasels that mate in summer, the reproductive cells lie inert for several months. Following this delayed implantation, the embryos develop quickly so that the young—usually three or four—will be born at the most propitious time of year. The three martens presented here all have the pale heads that typify the American species, though not the closely related European marten.

PLATE

41

Fisher

Martes pennanti

In some regions this large, darkly glossy species of marten is known by the Algonquian name *pekan,* and it is also called the black fox, Pennant's marten, and Pennant's cat. Indeed, the fisher looks like a cross between a pine marten and a fox or cat, and it shares traits of all three in its agility and speed, tireless hunting, buried caches, and taste for carrion. The origin of its most common name is unknown; Bachman theorized that early observers had mistakenly ascribed to it the fishing habits of the mink. Before the era of fur trapping and the clearing of vast woodlands, the fisher dwelt as far south as the Carolinas. Now it is a rare, fleeting shadow in Canadian forests and the northernmost timber of the United States. It is a solitary hunter of the deep woods, particularly spruce forests, where it patrols a circuit sometimes exceeding sixty miles. The young, usually two, are born in a tree hollow or rock den in April, and almost immediately the mother leaves them long enough to find a male and mate again. Close to a year will pass before she delivers her next cubs, because as with a number of other weasels, gestation is prolonged by delayed implantation. By then her previous cubs will be almost mature. Females tend to be hardly larger than pine martens but a male may be over three feet long and weigh fifteen pounds or more. The coat is gleaming brown, sometimes almost black, and frosted with white-tipped guard hairs. But such details vanish in a sinuous blur when a fisher is chasing a hare, flipping a porcupine over, or climbing after a squirrel.

No 9
PLATE XLI
Drawn from Nature by J. J. Audubon, F.R.S.F.L.S.
MUSTELA CANADENSIS, LINN.
PENNANTS MARTEN OR FISHER.
MALE.

Almost duplicating the pose Audubon
painted, a fisher begins to descend head-first from a
tree fork. If alarmed, it may beat its
forefeet against the trunk while clinging to the bark
with its hind claws. This drumming may be
a threat display, or as some observers suggest, a habit
limited to females—a warning to kits
hidden in a tree den. The marten above peers from such
a den. The nest, lined with grass or moss, is
located among rocks or in a log if no
tree cavity is available. In the summer, after three
months in the den, the young emerge and learn
to hunt. They do not leave their mother until winter.
When defending herself or her inquisitive and
inexperienced young, she will fight furiously; growling,
hissing, and screaming as she bites and claws.

PLATE

93

Black-footed Ferret

Mustela nigripes

The scarce and elusive black-footed ferret was first described by Audubon and Bachman in 1851 from a specimen procured by a fur trader in Montana. "It is with great pleasure that we introduce this handsome new species," they announced in the *Quadrupeds*. However, some critics suspected a hoax when almost a generation passed before another specimen was identified. This ferret of the Great Plains is a buff-yellow weasel and is one and a half to two feet long. It has a whitish face, a black mask across its eyes, black legs, and a black tail tip. Its discoverers correctly asserted that the animal "feeds on birds, small reptiles . . . eggs . . . hares," and similar prey, but they were unaware of its dependence on the prairie dog, its chief fare. Usurping a burrow after killing the occupant, it resides in a dog town, surrounded by food. A century of chemical warfare waged by stockmen and government exterminators has drastically reduced the population of prairie dogs and made the black-footed ferret one of America's rarest mammals. Restoration is being attempted by ecologists—perhaps just in time, perhaps too late.

PLATE XCIII.

After sitting erect to survey a
prairie-dog town, a black-footed ferret slinks
up to an occupied burrow. With its
body close to the ground, it slithers along like a
snake stalking a mouse until it can peer
into the entrance. If it sees its
prey near the surface, it will lunge; otherwise,
it can virtually flow down into the
hole. Having taken up residence in the burrow of
a victim, a ferret is sometimes
seen raising its black-masked head from the hole.
There, in late spring or summer, a
female bears her young—four or five as a rule.
If prairie dogs are scarce a ferret
will seek other prey, stalking or digging up mice,
ground squirrels, gophers, and the like.

PLATE

140
Long-tailed Weasel
Mustela frenata

"To us the Ermine, in its winter dress, has always appeared strikingly beautiful," wrote Audubon and Bachman, admiring its snowy coat and its black eyes turned to sapphires in the shifting light. They referred to the long-tailed species, though today the name ermine is chiefly applied to the short-tailed *Mustela erminea.* This plate by John Woodhouse Audubon was thought to represent yet another variety—the "little nimble weasel"—for the model was a long-tailed female half the size of a mature male, which may reach a length of two feet. Less northerly than the short-tailed weasel, the species thrives from Bolivia through most of Canada. The *Quadrupeds* revealed confusion over weasel classification because of the sexual difference in size as well as the seasonal molt. Only in northern latitudes do the animals whiten. In a marginal zone of about six hundred miles, some—predominantly females—turn white while others remain summer-brown. Thus, many are camouflaged for survival in mild or severe winters. An observer in Pennsylvania might find both brown weasels and white loping through ground cover, their heads craned for scent and backs arched to strike small prey from insects and mice to birds and rabbits.

PLATE CXI.
PUTORIUS AGILIS, AUD & BACH.
LITTLE NIMBLE WEASEL
MALE & FEMALE
Lith. Printed & Col. by J. T. Bowen, Philada. 1848

The spring molt turns a weasel skewbald as summer-brown replaces winter-white in spreading patches, like bits of earth exposed by the melting of snow. In autumn the same transitional mottling will occur—induced by the diminishing hours of daylight. Both the long-tailed weasel shown in three of these pictures and the more northerly short-tailed weasel, seen below in its winter coat, retain a black tail tip throughout the year.

No. 7.
Drawn from Nature by J. J. Audubon, F.R.S. F.L.S.
PUTORIUS VISON, LINN.
MINK.
Natural Size
MALE & FEMALE.

PLATE

33
Mink
Mustela vison

Relinquishing their hostile ways, two minks consort in early spring during their mating period and in summer when meat must be brought to the young in a streamside den. White-downed at birth, the kits—four or five, occasionally more—are weaned by then, active and hungry. The parents seek mice, birds, fish, rabbits, snakes, frogs, and almost anything they can kill. Prowling a stream within sight of a trapper's shanty, they are in little danger until late autumn, when the guard hairs of their coats acquire a sumptuous gleam. They are heavy-bodied but sinuous, the size of a marten, and dark brown except, as a rule, for a white chin and throat patch. The pastels of fashion are not usually bestowed by nature but by the selective breeding of mutations. So valuable is the fur that the species has been bred in captivity since the time of the Civil War. For every animal trapped, five pelts come from mink ranches. Living in the wild from the arctic almost to the Mexican border, the mink is rare only in arid country where there are no productive waterways to hunt.

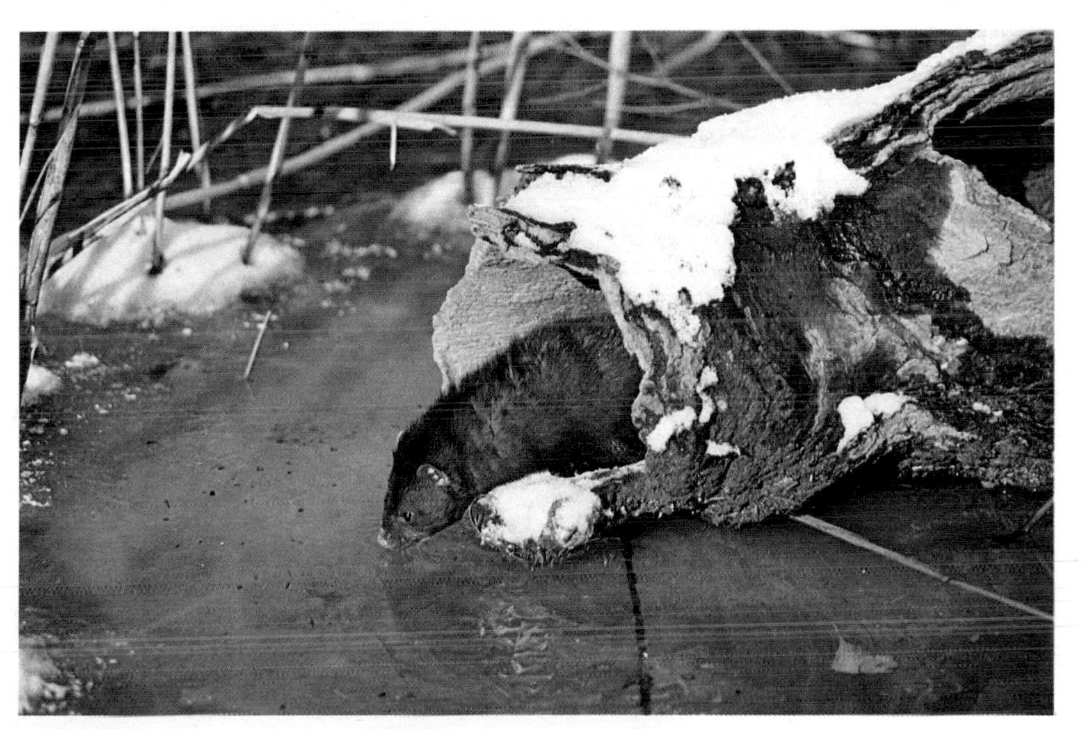

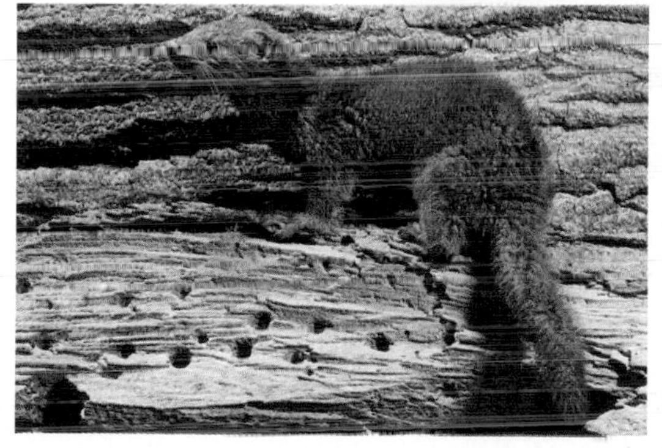

Audubon and Bachman emphasized the deft rapacity of the mink—"a cunning and destructive rogue" that kills poultry, catches rodents like a weasel and carries them off "by the neck in the manner of a cat," and is "a tolerably expert fisher." When angered or alarmed, a mink exudes a skunklike odor. With a lunge at the neck, it can slay animals larger than itself. Although it kills more than it can eat, its mood is not perpetually murderous. It will slide down banks like an otter and romp in a stream. Popping out of hollow logs, the juvenile at left and the adult above may be fishing or exercising. The one in a poplar may be intent on raiding a bird's nest or merely enjoying a climb.

Hunger or merely restless
curiosity can tempt the usually
nocturnal striped skunk to
scuffle about the woods in daylight,
as at right. Glistening
black and white fur cannot work very
effectively as camouflage,
but may warn predators off. Sometimes
before bracing its body
to spray an enemy, a skunk
lifts its hind legs and stands for an
instant on its front paws,
an action that may be a threat display.
Pictured above is a spotted skunk
(Spilogale putorius) digging
for prey—a mole, perhaps, or insects.
Though absent or scarce in
the North and along much of the
Eastern Seaboard, it is nearly as
widespread as the striped
species, ranging through Mexico and
most of the United States.
Its habits are essentially similar,
but it is smaller—seldom
much more than a foot long—and it is
the only tree-climbing skunk.

PLATE

47
Badger
Taxidea taxus

Bow-legged and squat as a marmot, a waddling badger may nonetheless seize an injured bird. It also devours eggs, nestlings, insects, and even snakes. Tough, loose hide and thick fur can resist a rattlesnake's venomous bite unless the face is struck. But rodents are the chief fare, and the grasslands are pitted with their excavated tunnels. They seldom escape, for powerful feet and very long front claws make the badger a remarkably fast, relentless digger. Across the western two thirds of the continent, north into lower

Canada, south into upper Mexico, this heavy, yellowish-gray weasel lives wherever abundant ground squirrels can be dug from a prairie. It has black feet, a white stripe over the head to the nose, and black-patched white cheeks. Unlike the Old World badger, it is not gregarious, and a fifteen-pound specimen can thrash most enemies. It does tolerate a foraging coyote that forces prey underground for the badger. The coyote then awaits its reward when a rodent eludes the badger's digging claws trying to escape through another hole.

A typically short-tailed, heavy-bodied, wide-shouldered badger, about two and a half feet long, is built for plowing into the loose soil of grasslands, as at far right. It digs for rodents, and although it is a fierce battler it prefers to evade larger predators by digging its way to safety. Bachman remarked of a captive male that he could "in a minute bury himself." Often the species seems to dig for the pleasure of digging. A female with young in her den returns regularly to feed them, but a badger with no young sometimes excavates a new burrow daily. The two or three kits, born in the spring, are nursed for a couple of months, after which the mother kills most of their prey for them until autumn, when they are nearly grown and ready to disband.

No 6.
Drawn from Nature by J.J. Audubon, F.R.S.F.L.S.
GULO LUSCUS, LIN.
THE WOLVERINE.
Natural Size

PLATE

26
Wolverine
Gulo luscus

Largest and cleverest of weasels, the wolverine is also called the glutton (a literal translation of its generic designation) and the skunk-bear. It has a glutton's appetite, a skunk's odor, and a shape vaguely like a bear in miniature. Usually about three feet long and a bit over a foot high and weighing up to forty pounds, this shaggy, big-footed, short-tailed marauder wanders the forests and tundra where its bear-sized antecedents roamed four million years ago. Its coat is a rich brown, often blackish, with a lighter band along the flanks and over the rump. The circumpolar range of the species has been reduced in North America to Alaska, northern and western Canada, and the high Sierra Nevadas. Eating anything it can vanquish, it subsists chiefly on small prey, carrion, and berries, but will drive larger predators from their kills and will bring down a big-game animal that is disabled or hampered by deep snow. If woodsmen respect its cunning, they fear its rage when cornered and detest its seemingly malicious urge to destroy. It will follow a trap line to steal the bait or eat any trapped animals, and foul or tear apart any cached provisions it cannot eat. At the same time it protects its own caches with a musk that discourages thievery by other carnivores.

The bearish look of a wolverine's broad
head and rounded back is vivid in two of these portraits,
yet the hulking body is equally reminiscent
of a badger's. Audubon and Bachman reported that this
strange boreal recluse had "always existed
very sparingly" in the United States, "and only in the
Northern districts." Bachman once came upon a
wolverine's trail while hunting hares
near Albany, New York. Supposing the tracks to be those
of a small bear, he followed them through the
snow to a rocky cavern. When he probed the dark recess
with a pole, the growling occupant bit
it and tore it from his grasp. Finally the animal
was killed and examined. Bachman believed
he had discovered a new species, but his elation subsided
six months later when he found, upon consulting
Georges Louis Buffon's *Histoire Naturelle,* that his curious
specimen "was the Glutton, of which we had read
such marvellous tales in the school-books."

PLATE

21
Gray Fox
Urocyon cinereoargenteus

Catching the tantalizing scent of down fluttering overhead, a loping fox crouches and tilts its nose skyward. Bachman and Audubon said that when the gray fox hunts birds it "winds them like a pointer dog." This one has flushed a duck, goose, or partridge; or the feather may have wafted up from poultry at a farm in the valley. But the gray fox, unlike the red, will hesitate to raid a farm unless driven by prolonged hunger. It prefers to seek rodents, rabbits, carrion, and fruits such as cherries and grapes in the security of the surrounding woods. Found from Central America up through the United States (though rare in the Midwest and upper West), it is the only tree-climbing American canine. It is subtly unique in appearance as well as behavior. As heavy as a large domestic cat, it tends to be slightly smaller than the red fox. Its salt-and-pepper coat, washed with rust on flanks, legs, neck, and sometimes elsewhere, can be hard to tell from that of the red variety, but the top of the tail has a black streak and is always tipped with black, whereas the tail of a red fox is white-tipped.

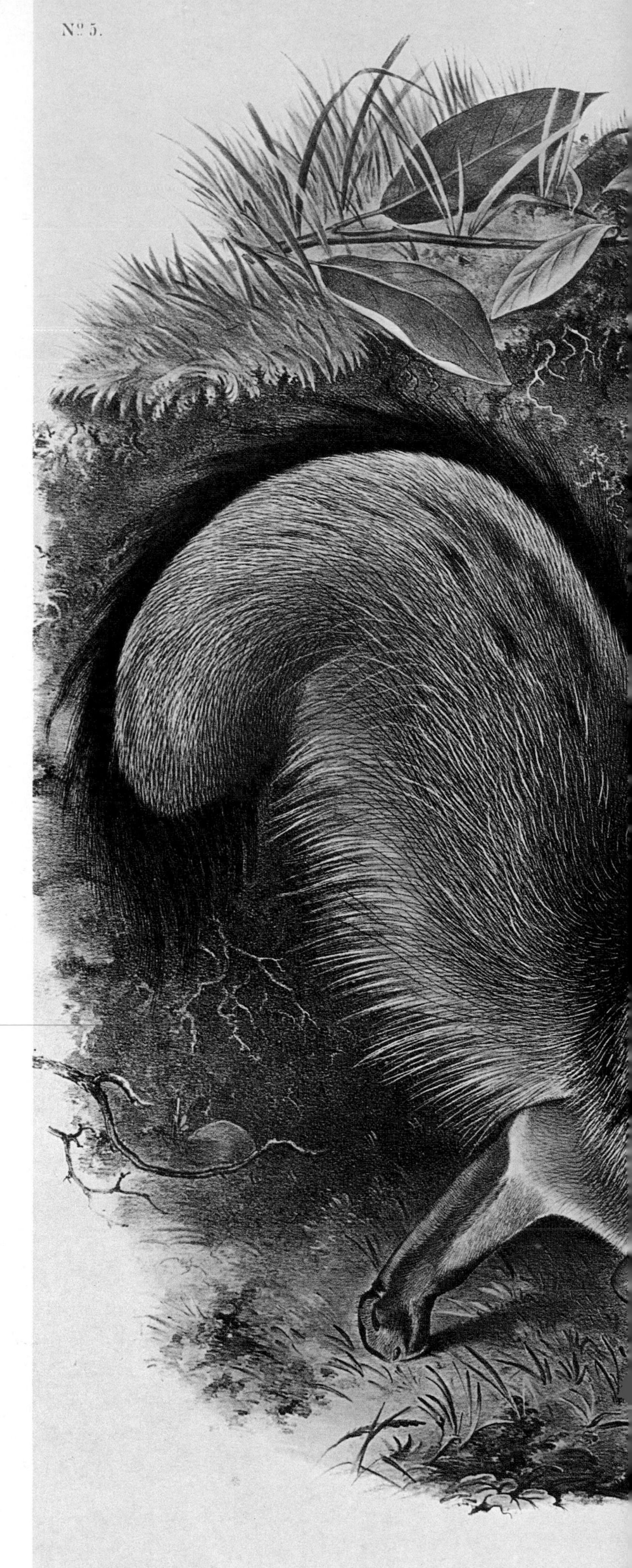

Nº 5.
Drawn from Nature by J. J. Audubon, F.R.S. F.L.S.

PLATE XXI.

CANIS (VULPES) VIRGINIANUS, GMEL.

GREY FOX

Natural Size

Lith. Printed & Col.d by J. T. Bowen, Philad.a 1843.

A red fox looks back at a snow-blanketed mousing field from a brushy hillock. Having eaten well and perhaps cached a surplus of prey or carrion, it is about to bed down on some little rise where it will doze fitfully through the morning, raising its head frequently to be sure no enemy has trailed it. A red fox may take shelter in a storm but usually disdains a den until breeding season. Catnapping in the open, it curls up with its brushy tail over nose and feet for insulation. In January or February it mates, and the female then selects a den site that the male helps her to prepare for whelping. They may dig a burrow or use a natural cavity, but red foxes prefer to enlarge a marmot burrow. If a hibernating marmot is in residence, they have secured food as well as a den. The kits, averaging five, are usually born in March or April. Two months later, when the kits are weaned, the parents begin to drop food farther and farther from the den, enticing them to explore. By late August the young have been taught to hunt, and the family will shortly disperse.

No 15.

Drawn from Nature by J. J. Audubon F.R.S.F.L.S.

CANIS LATRANS, SAY.
PRAIRIE WOLF.
MALES.

PLATE

71

Coyote

Canis latrans

"We saw a good number of these small wolves on our trip up the Missouri," Audubon reported. "This species . . . has much the appearance of the common grey Wolf . . . but differs from it in size and manners. . . . The bark or howl . . . resembles that of the dog, and on one occasion the party travelling with us were impressed by the idea that Indians were in our vicinity, as a great many of these wolves . . . barked during the night like Indian dogs. We were all on the alert, and our guns were loaded with ball in readiness for an attack." The coyote is the smallest American wolf, barely two feet high and three feet long, with another foot or so of brushy tail swaying behind. Audubon noted that it digs a family den like that of the red fox and hunts in packs—family groups. The half-dozen young usually disperse after their first summer, but mates may stay together for several years or for life. So indiscriminate is their appetite that one zoologist speculated they have little sense of taste. Coyotes, often called prairie wolves, have defied ranchers' misguided efforts at extermination and in recent years have multiplied and spread into the eastern woodlands.

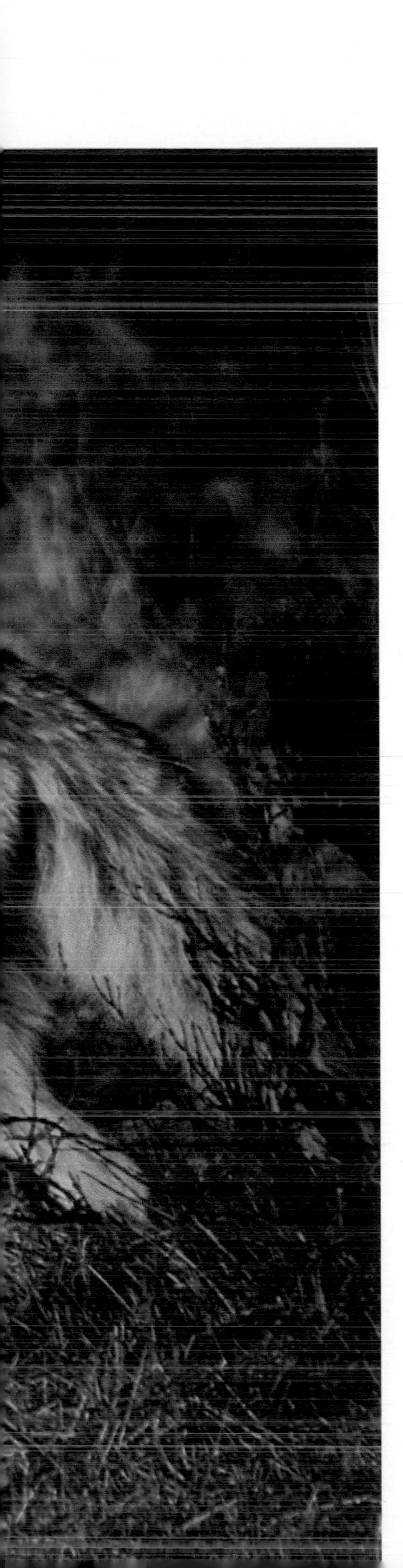

With a forepaw lifted in readiness to flee or pounce, a coyote resembles a nervously alert domestic dog except for its slanted lupine eyes. It is probably hunting rodents by scent and sound. Its rough coat blends with prairie grass, brush, or desert sand, yet it may be sharply silhouetted against the sky when it stands on a ridge to scan the locale for hares and the like. Since it cannot outrun a jack rabbit or a pronghorn fawn that is more than a few weeks old, two coyotes may run in relays to tire their prey, or a coyote may herd a victim toward a partner lying in ambush. Members of a family pack signal one another by howling and yapping, as pictured above. At right, a pup emerges from a den. In a few more weeks—before the onset of summer—the den will be abandoned and the young will follow as the parents wander and hunt.

PLATE

82
Red Wolf
Canis rufus

So variable is the color of this wild dog that sometimes it is called the black wolf rather than the red wolf. It looks like a cross between the larger gray wolf and the smaller coyote. Indeed, it interbreeds readily with the coyote, and some of the remaining packs are hybridized. The ventriloquial howls of red wolves once awed settlers from Illinois to Texas and from the Carolinas to Florida. Only a few packs survive along the Gulf coast of Texas and in Louisiana, and fewer still exist in isolated pockets of habitat in Mississippi, Arkansas, and Oklahoma. Yet even now conservationists must persuade some ranchers that the species does not threaten their stock and should not be exterminated. Red wolves chase young deer but subsist mainly on rabbits, hares, rodents, and ground-nesting birds. On shorelines they will also eat crustaceans. They are very bold scavengers—in Florida they stole a string of fish from the camp of the naturalist William Bartram, and in Texas a scout told John Woodhouse Audubon of having his roasting venison pulled from the spit by a red wolf.

PLATE LXXXII.

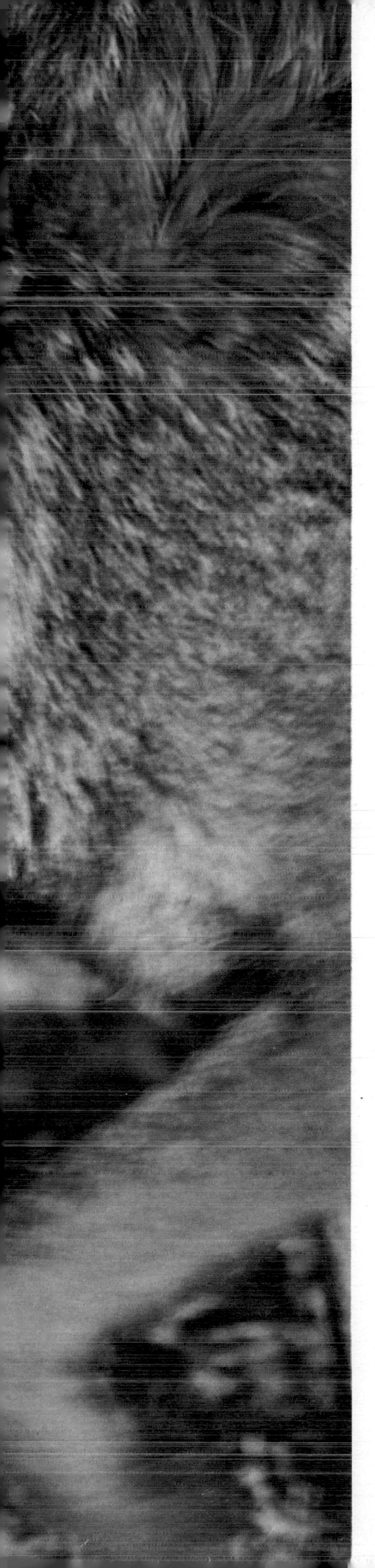

Larger and more heavily muscled than a
coyote but leaner and smaller than a gray wolf, the red
wolf is built for arduous prowls through swamps
and bayous and long, loping runs over
coastal prairies. The male at left probably weighs
between fifty and seventy-five pounds,
twenty percent more than a female. Before setting off
to hunt in the late afternoon or evening,
wolves often bark and howl in chorus, strengthening the
pack's bonds and perhaps whetting an appetite
for pursuit. Above, a female carries a woolly brown pup
away from a den that has been disturbed.
There may be two or three others in the litter or nearly
a dozen. They weigh hardly a pound each at
birth in the early spring, but by fall they are big
enough to travel with their elders.

No. 14.

Mark Catesby, one of the first chroniclers of American wildlife, wrote in 1743 that wolves "go in droves . . . with dismal yelling cries." Audubon, too, observed large howling packs as well as lone wolves before much of the wilderness was converted to farms, ranches, and towns. Traveling up the Missouri River he noted that "some days . . . we saw from twelve to twenty-five." Years before in Kentucky he had seen an abundance of black or nearly black specimens like

the one depicted here. Yet in the West he found most o them to be gray, brindled, or white. He painted both the black and white extremes but realized that the were mere "varieties" of the same "bold and savage" hunter of the timber and prairie. Some wolves, chiefly in the Arctic, are pure white, and some tundra wolves weigh well over one hundred pounds. Mo typical, however, is an eighty-pound gray or tannish-gray timber wolf, resembling a very large

PLATE LXVII.

PLATE

67

Gray Wolf

Canis lupus

German shepherd dog. Family packs of such
wolves—mates, siblings, and a half dozen or so young—
once rendezvoused at dens and caches almost
everywhere in North America. Healthy populations remain
in Alaska and Canada, but below the Canadian
border the last of them are concentrated in Minnesota's
woods. Atoning for man's past ecological sins,
wildlife specialists are trying to restore the wolf to
its former habitat in other suitable regions.

A timber wolf savors a kill that
may have been brought down by the mass assault of
a pack or an explosive lunge by the
single animal now presiding over the feast. A
lone wolf can fell a winter-weakened
deer, though seldom by tiring it in the harrowing
pursuits described in legends. Wolves are
too intelligent to squander energy
very often in long chases of dubious outcome.
They rely more often on a sudden,
fairly short dash whose success depends on
surprise, the craftiness of a single wolf or the
teamwork of a pack, and the condition
of the prey. Much of the carcass at right has
been devoured. If prey is abundant
the remainder may be abandoned, but wolves
frequently cache meat for future use. The feeding
wolves directly below were found
in the northern Rockies; the one at right and the
pack laboring through deep snow were
found in the Yukon. They show the pronounced
color variations that are common
among nominally gray wolves.

No 12.
BOS AMERICANUS, GMEL
AMERICAN BISON OR BUFFALO.
MALE.

PLATE

56 Bison

Bison bison

Near the Yellowstone on August 19, 1843, Audubon wrote that the wolves' howling and bison's roaring sounded "like the long continued roll of a hundred drums." Two days later an entry described "buffaloes all over the bars and prairies, and many swimming; the roaring can be heard for miles." He did not exaggerate. There were thousands, and a large bull, weighing nearly a ton, bellows mightily during summer's harem-gathering frenzy. His account of breeding was incorrect when he failed to recognize that the clashing battles he witnessed marked the peak of mating rivalry. He was more accurate in recalling that during his youth, bison had dotted the "beautiful prairies of Indiana and Illinois, and herds . . . stalked through the open woods of Kentucky and Tennessee." From Canada to Mexico and almost from coast to coast, at least thirty million grazed. By 1900, after the Indians had been starved, the railroad builders fed, and the fashionable East robed, fewer than a thousand remained. Then an unprecedented conservation movement arose, and despite severely reduced habitat small herds were rescued. Today more than thirty thousand bison have been restored to the western grasslands.

A rutting bull bison deters rivals by lowering his head, pawing the ground, lolling his tongue, bellowing, and raising his tail. If threat displays fail, butting and goring may ensue. The beast's powerful neck, skull structure, and thick boss of hair on the forehead absorb shock. Dominant bulls tend to have large mats of hair and copious beards. Buffalo roll to get rid of insects and relieve itching, but wallowing has another function in the rut. A threatened bull can signal submission and avoid a fight by what ethologists call a displacement activity. This often takes the form of dusting, which dissipates frustration and tension. Calves are born in the spring and usually stay in the herd, near their mothers, until they are three years old. Cows first breed at that age, and four-year-old bulls compete for them, but males and females alike continue to grow for several more years.

No 16
ANTILOPE AMERICANA, ORD.
PRONG-HORNED ANTELOPE.
MALES & FEMALE

PLATE LXXVII

PLATE

77
Pronghorn

Antilocapra americana

This uniquely American creature, popularly called the pronghorned antelope, is the only living species of its genus. It is not an antelope nor does it have any close relatives, though it shares anatomical details with antelopes, goats, and giraffes. A mature buck stands about three feet high at the shoulder and weighs hardly more than one hundred pounds—a deceptively fragile-looking animal with leg bones stronger than a domestic bull's and a long-distance running speed unmatched by any beast. A band of pronghorns can lope along at thirty miles an hour for thirty minutes and sprint for more than a mile at twice that speed. "Their walk," Audubon remarked, "is a slow and somewhat pompous gait, their trot elegant and graceful, and their gallop . . . light and inconceivably swift." He also observed—and demonstrated to skeptical hunters at Fort Union—that the species neither keeps its horns permanently like other horned animals nor sheds them entirely like antlered species but drops the sheaths and retains the hard membranous cores. Females usually have horns, but they are small, prongless spikes or bumps, never the heavily curving black lyre adorning a buck's head. Pronghorn herds, grazing in bands of half a dozen to a score or more, have been fenced out of vast cattle-grazing tracts but they still inhabit the plains from southern Canada to Mexico and from the Dakotas almost to the Pacific.

A complex musculature enables the pronghorn to erect the long hairs of its bright white rump patch. When alarmed, it can flare this rosette out for several inches, flashing a signal like a heliograph to all pronghorns within sight. Their keen eyes can detect the alarm signal across miles of flat prairie or desert. A mature buck like the one pictured usually lags behind a fleeing band, but not for lack of speed. It can run fifteen miles without resting. A mature doe generally takes the lead while a dominant buck acts as a rear guard, herding indecisive does and fawns. Birth occurs in the spring, and a first pregnancy is apt to bring a single fawn, after which twins are the rule. Nearly odorless for the first few days, they lie quietly in tall grass or brush while the mother feeds at a distance so that she will draw no enemies to them. But within a fortnight they follow her confidently, for by then they can outrun most predators.

PLATE

128

Mountain Goat

Oreamnos americanus

Having crossed the Bering land bridge from Asia more than half a million years ago, mountain goats now exist only in North America. They browse and graze amid the crags and slopes of southeastern Alaska, the Yukon, western Canada, down through the coastal ranges into Oregon, and through the Rockies into Wyoming. The species is not a true goat but a creamy-white antelope related to the European chamois and the Himalayan serow. Slab-sided, hump-shouldered, and bearded, it has almost the contours of a miniature bison except for the curving black stiletto horns eight or nine inches long that are atop its rectangular head. A nanny rarely weighs two hundred pounds; a large billy may weigh three hundred. Its shaggy winter pelage makes the animal appear larger until it sheds in tufts and tatters, leaving a sleeker summer coat. A mountain goat threads the narrowest ledges and surveys its world from seemingly inaccessible pinnacles. Its double-lobed hoofs are perfectly adapted for rock-climbing—their slightly convex, rubbery soles providing traction while the hard rims hook the smallest projections. A nanny's single kid, born in the spring, is soon sure-footed enough to follow its mother as she wanders the highest, steepest pastures.

PLATE CXXVIII.
BLAINVILLE.
ROCKY MOUNTAIN GOAT.

"Fearless climber of the steeps," Audubon and Bachman called the mountain goat, admiring the way it "springs with great activity from crag to crag" and is seen "browsing on the extreme verge of some perpendicular rock." Of the three at left, one appears to be a kid or small yearling, already as agile as its elders. A tattered-looking companion, not yet fully shed of its winter coat, is probably the dam. She stands on so small a protuberance that she seems to defy gravity. A third occupies a small ledge while nibbling at invisible rock lichens. The nanny pictured below, with her kid, is in her relatively thin summer pelage; the billy at right is garbed for winter, and the one at right below was photographed in the early spring.

N°15.

OVIS MONTANA, DESM

ROCKY MOUNTAIN SHEEP

PLATE

73

Bighorn Sheep

Ovis canadensis

Pictured here are a Rocky Mountain bighorn ram and ewe of the Badlands subspecies, now extinct but plentiful when Audubon explored eastern Montana. Of the many such localized races there are four major varieties but only two species of North American wild sheep. Both the Rocky Mountain bighorn and the leaner, paler desert bighorn are *Ovis canadensis*, a species ranging from British Columbia's Rockies southward into Mexico and eastward into the Dakotas. The white Dall sheep and the nearly black Stone sheep are both O. *dalli*; the white type occupies Alaska and far northwestern Canada while the overlapping range of the dark Stone lies slightly to the south. Largest of the group is the brown, pale-rumped bighorn sheep of the Rockies; a ram may weigh three hundred pounds and carry massive horns that sweep around in a full curl of forty inches or more. The skimpily horned ewe is smaller but she is the male's equal as a mountaineer. Sheep do not pick their way like mountain goats but bound from point to point on sharp-rimmed, spongy, slightly concave hoofs. During the autumn rut, males are literally battering rams as they duel for the scattered bands of ewes. The head-on clash of horns often reverberates for more than a mile across talus slopes and through the echoing canyons.

On the watch for predators, the amber
and umber eye of a desert bighorn gazes down with
assurance on an arid or mountainous
realm. Its vision is said to equal that of a man
aided by an eight-power binocular. One
of the sheep's massive, curling horns is "broomed"
—broken or rubbed away at the tip.
Only occasionally does this happen by accident.
An old ram like the one pictured
files his horns off against rocks when the
tips begin to block his peripheral vision. At top
is a young Rocky Mountain ram, not
yet two years old; his horns, larger than a ewe's,
will grow for at least five more years.
Lambs are dropped in the spring
and they remain with the bands of ewes for a
couple of years before the young males
form bachelor bands of their own. Above, two
mature rams crash head on at more
than twenty miles an hour, producing over a ton
of impact as they fight for ewes.

PLATE

136

White-tailed Deer

Odocoileus virginianus virginianus

America's most adaptable hoofed animal, the white-tailed deer flourishes from the southern half of Canada down through Central America. It is rare or absent only in small portions of the southwestern United States. Because brush and second-growth woods provide ideal browse and cover, the species has profited from lumbering and the abandonment of small farms, and is more abundant now than when the continent was discovered. An average buck weighs a hundred and fifty pounds and a doe somewhat less. But among the thirty subspecies are northeastern and midwestern deer that occasionally weigh twice as much, while in the Florida Keys and on Panama's Coiba Island the bucks are no larger than a lean collie dog. In addition to the finely antlered eastern buck and graceful doe presented here, Audubon painted a fawn in the white-spotted coat of its first summer and a buck of the Columbian subspecies (*O. v. leucurus*) which he mistook for a distinct species and labeled "long-tailed deer." The tail, outwardly brownish, is white on the fringe and inner surface. When startled, the deer tosses its rump high and flips its tail up in a flashing white alarm signal as it bounds into the concealing forest.

N° 28.

Drawn from Nature by J.W. Audubon

PLATE CXXXVI.

CERVUS VIRGINIANUS, PENNANT.

COMMON OR VIRGINIAN DEER.

Lith.d Printed & Col.d by J. T. Bowen, Philad.a 1848

On a clear winter morning a
four- or five-year-old whitetail buck picks
his way over thin snow from browsing
grounds to a higher, more concealed bedding
area. Soon after the rut subsides,
he will shed his long-tined antlers. In May,
when he begins to grow new ones,
many does will bear their young—single fawns
after a first mating, usually twins and
occasionally triplets thereafter.
A newborn fawn weighs no more than a human
infant, but by late autumn some
fawns are almost as big as their mothers.
In summer, a white-spotted russet fawn blends
into sun-dappled thickets. The
spots, often as many as three hundred, fade
as the animal develops its first
grayish-brown winter coat.

PLATE

78
Mule Deer
Odocoileus hemionus hemionus

Audubon explored the West in spring and summer while the half-developed antlers of buck mule deer were sheathed in the velvety membranes that supply blood to the growing bone. "We have figured a female in summer pelage," he wrote, "and have represented the animal in an exhausted state, wounded . . . about to drop . . . whilst the hunter is seen approaching, through the tall grass." Evidently he would have preferred to depict a regally crowned buck but the expedition's hunters brought him one with small antlers. Instead, a plump doe was shot and used as a model before becoming table fare. Her "summer pelage" would have turned grayish by winter and she would have grown fatter on an assortmen of western browse. In October or November she would have mated; with the deepening of snows she and a group of companions would have left the high plains and plateaus for more sheltered valleys; and in the spring she probably would have dropped twin spotted fawns. Several races of mule deer range from southeastern Alaska (where the species is represented by the Sitka blacktail, *O. h. sitkensis*) through western Canada, southward into Mexico, and from the Pacific to the Midwest. All have big ears and black-tipped tails, but some along the coast and in the deserts are built like white-tailed deer while others, particularly in the upper Rockies and on the plains, grow considerably larger.

PLATE LXXVIII.

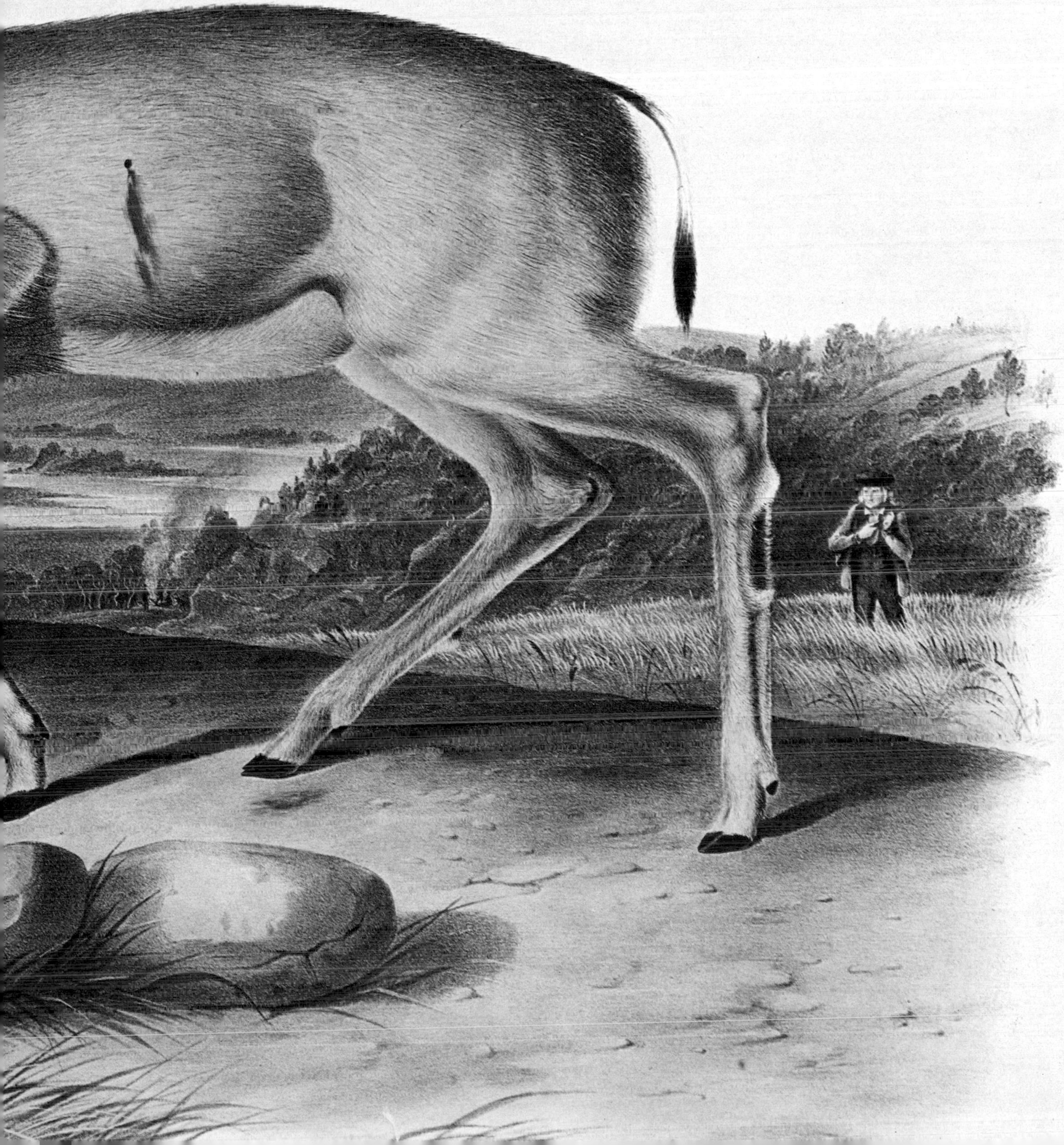

A buck mule deer's antlers curve gracefully
out and upward to a fork, and each resulting branch forks again,
like the crotches of a growing tree. Thus the
antlers help to distinguish the species from the whitetail,
whose antlers typically consist of unforked
beams, sweeping forward more or less horizontally, with
tines rising from them. A mule deer's roughly cylindrical tail
is much narrower than that of the white-tailed deer;
it actually shows far more white on the outer surface but is not
raised in alarm to display the white underside,
in the manner of the whitetail. A running mule deer tucks
its tail in and bounces along, then pauses to look
back and ascertain whether the source of its
fright—real or imagined—is gone.

CERVUS RICHARDSONII, AUD & BACH.
COLUMBIAN BLACK-TAILED DEER.
Male.

PLATE

106

Columbian Black-tailed Deer

Odocoileus hemionus columbianus

Lewis and Clark described this race of mule deer, which they encountered near the mouth of the Columbia, but as the text of the *Quadrupeds* stated "not until the discovery of the golden treasures of California did it become generally known to white men." The meat of this and other game brought such high prices at the diggings that many miners turned from prospecting to professional hunting. The Columbian blacktail is found in the rain forests of coastal British Columbia and southward into the arid canyons and rasping brush of central California. Whereas an inland mule deer has a roughly cylindrical white tail with a black tip, this deer has a broader tail with a brown or blackish outer surface down to the black tip. The Columbian blacktail is a trifle leggier than the more northerly Sitka blacktail, but both are the size of white-tailed deer, with antlers less massive than those of other mule deer. In Audubon's time and long afterward blacktails were regarded as a separate species. However, their antler formation and other characteristics are basically those of O. *hemionus,* with which they interbreed where the ranges overlap, thus proving that the two coastal blacktails are subspecies of the inland mule deer.

A young blacktail buck at first glance
resembles a doe as he ambles through the grasses and wild flowers
of a mountain pasture near the Pacific coast.
Later in the summer his velvet-cased antlers will begin to
fork, but now they are not yet as long as
his ears. Below, a doe lingers near a snow patch where the air
is relatively cool and free of annoying insects.
The bottom picture shows a mature buck during the autumn rut. Both
the Columbian black-tailed subspecies and the Sitka
blacktail *(O. h. sitkensis)* look much like other mule deer, but
their tails are shorter, wider, bushier, and blackish or
brown on the outer surface down to the black tip.

PLATE

62

Elk

Cervus canadensis

By the 1840's when Audubon penned his notes for the *Quadrupeds,* elk had become rare in the East. He had seen them in Kentucky only three decades earlier, and the models for his plate were a bull and cow captured in Pennsylvania and kept at his New York estate. After encountering the species again on the Missouri, he described a rutting bull: "His spreading antlers have acquired their full growth, the velvet has been rubbed off, and they are hard and polished. . . . He stamps the earth . . . and utters a shrill cry somewhat like the noise made by the loon. When he discovers a group of females he raises his head . . . and giving another trumpet-like whistle, dashes off to meet them, making the willows and other small trees yield and crack." A large bull is the size of a horse and his antlers alone may weigh fifty pounds. Early settlers called this awesome deer an elk, their name for the European moose; later, the Shawnee name "wapiti" also gained currency. Today the several remaining races of elk, or wapiti, range chiefly through the Rockies, the Cascade and Olympic mountains, and the lower parts of Canada's Prairie Provinces.

PLATE LXII.
CERVUS CANADENSIS, RAY.
AMERICAN ELK — WAPITI DEER.
MALE AND FEMALE.
Lith. Printed & Col.d by J. T. Bowen, Philad.a 1845

Alertly guarding his cows, a bull elk of the Rocky Mountain race (*C. c. nelsoni*) turns his head as he hears the shrill bugle of an unattached male seeking a herd that can be commandeered. The six long tines on each enormous beam of his antlers mark him as mature, at least four years old. The tensions and hostilities of the autumn mating season have not yet enervated him, and he probably weighs over seven hundred pounds. He may reply to the challenge by lifting his muzzle and uttering a warning bugle of his own, as in the picture at near right. In May or June, a typical cow will bear a single, white-speckled, thirty-pound calf. Above, a bull of the heavy-bodied Roosevelt, or Olympic, race (*C. c. roosevelti*) near the Pacific coast lords a harem of more than a dozen cows after he has driven their yearlings away.

PLATE

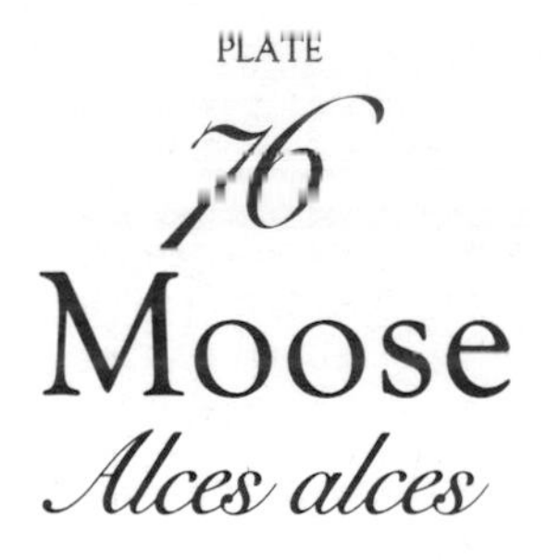

76

Moose

Alces alces

The world's largest deer, a bull moose is crowned in autumn with palmated antlers that may spread more than six feet across. The huge Alaskan subspecies is apt to stand seven feet high at the shoulder and pock the muskeg with cloven tracks pressed deep by a weight of perhaps seventeen hundred pounds. Cows are smaller, yet a cow of the least impressive American race, the Wyoming subspecies, may weigh six hundred pounds. Seven races of the single species occur in northern forests around the world. American moose range from Newfoundland through Canada and Alaska. A few dwell in the Maine woods, more in Minnesota, and still more in the Rockies south to Colorado. In a day, one of these enormous beasts can eat sixty pounds of shrubs, grasses, aquatic plants, tree bark, leaves, and twigs. The flexible, pendulous nose and upper lip peel away tough browse; less certain is the function of the bearded dewlap hanging from the throat, but it may drain off water as a wading moose lifts its head after nibbling lilies or drinking. Though rutting bulls are famous for aggression, it is probably safer to encounter a bull in autumn than a cow protecting her awkward, leggy calf in spring or early summer.

PLATE LXXVI

CERVUS ALCES, LINN

MOOSE DEER.

OLD MALE & YOUNG

One of Audubon's correspondents in Quebec, relating the habits of moose, reported that "in the summer they are fond of frequenting lakes and rivers, not only to escape the attacks of insects which then molest them, but also to avoid injuring their antlers, which during their growth are very soft . . . and besides, such situations afford them abundance of food." At left, a bull of the largest subspecies, the Alaska-Yukon race *(A. a. gigas)*, tends a cow during the rut. His antlers may weigh more than eighty-five pounds, yet they do not appear large in relation to his head and body. Above is a representative of the Wyoming subspecies *(A. a. shirasi)* with his antlers in velvet. Even a bull of this smallest race may stand six feet high at the withers. In September, he is likely to stay with a cow for the week or ten days of her estrus and then find another mate. He will tolerate the presence of a calf if she has one, but the cow herself will drive it off eight months later, before she bears another.

No 26.
Drawn from Nature by J. W. Audubon
TARANDUS FURCIFER, AGASSIZ.
CARIBOU OR AMERICAN REIN-DEER
MALES.
1 Summer pelage 2 winter pelage

PLATE CXXVI.

Lith.d Printed & Col.d by J.T.Bowen, Philad.a 1847

PLATE

126

Caribou

Rangifer tarandus

Precious wild beef to Eskimos and northern Indians, the caribou belongs to the same species as the more readily domesticated European and Asian reindeer. The dozen American subspecies are of three main types—the magnificently crowned but small-bodied barren-ground caribou of the tundra and taiga from Labrador to Alaska; British Columbia's mountain caribou, lighter-antlered but almost as big as elk; and the fairly heavy-bodied woodland caribou, whose shorter, narrower horn structure facilitates movement through timber and brush of a slightly more southerly range across Canada. The horns of caribou cows, the only antlered female deer, never rival the baroquely variable, widely palmated, multitined racks of the bulls. The plate shows two woodland males, one in summer velvet, the other with the copious whitish mane that adorns a fully mature bull. Some caribou herds on the barren grounds migrate eight hundred miles in the fall, leaving the shrubs and pale lichens of the tundra for the browse of forest edges. Big, splayed hoofs provide sure footing on ice, snow, or the spongiest bog, and their passage creates a unique sound, as long ankle tendons click with each step.

An Alaskan barren-ground caribou *(R. t. stonei)* grazes on the rank summer foods of the tundra—grasses and herbs as well as lichens, fungi, dwarf willow and birch, berry bushes, and the like. This bull is not very old, but he is over four feet high at the shoulder and his antlers sweep three feet above his head. In late August, when the rut begins, the antlers are hard though still in velvet. Bulls in some of the barren-ground herds do not gather harems but mate with any receptive cows. Among other barren-ground herds as well as woodland and mountain caribou, bulls defend harems of half a dozen to a dozen. Single births are the rule, but cows of the woodland subspecies *(R. t. caribou)* produce twins fairly often. Calving reaches a peak in June, en route to the summer meadows. "During our expeditions in Labrador," Audubon wrote, "we saw many trails [illegible] about as broad as a cowpath, and many times the fatigues of a long day's hunt over the sterile wilds of that country were lightened by following in these tracks . . . instead of walking on the yielding moss."

Drawn from Nature by J. J. Audubon, F.R.S. F.L.S.

DYCOTELES TORQUATUS, F. CUV.

COLLARED PECCARY.

MALE.

PLATE. XXXI.

PLATE

31

Collared Peccary

Tayassu tajacu

A band of a dozen or two dozen peccaries looks like a flock of dwarfed wild boar, rooting up tubers with their tough, flat nose pads, avidly crunching the thorniest prickly pear, snuffing about for mesquite beans, scrub-oak acorns, berries, and the like, or lunging at small prey. A barrel-bodied adult weighs forty pounds or so. It has a dark, bristly, peppery-gray coat dusted with tan. Usually a whitish band extends from the throat over the shoulders, but in some regions the collar is absent. Although related to true swine, the species belongs to a separate family. Its tusks are shorter and straighter than the wild boar's, and it has a dorsal musk gland forward of the rump. Indeed, some southwesterners call it a musk hog. A more popular name is javelina, a corruption of the Spanish word *jabali*—meaning boar. It is found in New Mexico, Arizona, Texas, and Latin America. Before its numbers were decimated for meat, "pigskin," and to protect crops, the peccary ranged into Arkansas. Because Audubon did not observe the species in the wild, his description perpetuated tales of a ferocity and resourcefulness typifying Europe's wild boar rather than America's small and timid peccary.

With mane bristling and rump hair raised as its dorsal gland releases a warning musk, a peccary dodges away from danger. A band of peccaries may number only three or four, or two dozen. Scattering when startled and regrouping by scent and sound, they seldom run far but their short legs can take them over steep foothills at twenty-five miles an hour. Though a collared peccary uses its hoofs and short, sharp tusks to fight furiously if cornered, it is less aggressive than the slightly larger white-lipped peccary *(T. pecari)* of South America. Both species evolved from giant piglike creatures that lived twenty-five million years ago. They have fewer teeth and a more complex stomach than true swine, and each hind foot has one dewclaw rather than two. Mating intermittently and polygamously throughout the year, they commonly produce twins, but some sows bear triplets or a single piglet.

Nº 20.
Drawn from Nature by J. W. Audubon.
FELIS CONCOLOR, LINN.
THE COUGAR
FEMALE & YOUNG.

PLATE

97
Cougar
Felis concolor

A mature male cougar usually weighs about a hundred and forty pounds and measures almost nine feet from its nose to the dark tip of its two-and-a-half-foot tail. Even a female, though lighter, stands over two feet high at the shoulder. But in spite of its size the species is rarely seen as it prowls through forests and rocky canyons; it is among the stealthiest of predators. Known by many names—mountain lion, puma, panther, catamount—it is an exclusively American feline that once ranged from coast to coast. When neither raising cubs nor mating, it goes its way alone and must have a large, game-rich hunting territory. Most of its eastern habitat has been usurped by man, and it has been obliterated in many areas where it was regarded as a potential menace to livestock. Cougars still inhabit the west from British Columbia to Patagonia, but only a few exist east of the Rockies—along the Gulf and in Florida. Except for a very dark Floridian subspecies, all the localized races are as tawny as deer, their favorite prey. The young, most commonly numbering two or three, are born in warm weather and hunt with the mother for at least their first year. Audubon painted a lioness with a half-grown cub whose infant camouflage of spots and short stripes had not quite faded.

The traditionally feared cougar,
Audubon and Bachman correctly asserted, is a
man-shy predator, very rarely willing
to stalk our species. A cougar's face seems to
evince both patience and restlessness,
tension and repose. The cat prowls, as the text
of the *Quadrupeds* declared, "with
a silent, cautious step, and with great patience
makes its noiseless way through the
tangled thickets." Yet this patient carnivore is
too restless to wait long in ambush,
crouching in brush or draped on a boulder
overlooking a game trail. Soon
it rises, stretches, and resumes a patrol
of its home range, which may cover thirty square
miles. Occasionally it surprises a
rabbit, rodent, or other small prey, but since it
can outrun a deer for only a short distance
it relies on stealth and a prodigious
leaping ability. Eventually it scents food, or
sees a glimmer of movement, or hears
the rustle of browsing. With its belly almost
touching the ground, it slinks close—
perhaps to within thirty feet—then pounces.

PLATE

1
Bobcat
Lynx rufus

The bobtailed American wildcat is a patient, usually nocturnal seeker of rabbits and other small prey, locating food more often by sight and sound than by scent and pinning it with cool efficiency after a short stalk and a pounce. But a bobcat that happens on a weakened deer or unwary fawn is twenty pounds of blurred fury streaking for the quarry's throat. A bobcat will not attack a man, but this painting evokes the hissing, crouching defiance of the animal when treed or bayed. A cornered bobcat will battle a pack of dogs with a savagery that inspired a frontier compliment reserved for the toughest of brawlers: "He can whip his weight in wildcats." In addition to this eastern specimen, Audubon painted a smaller, lighter-coated Texas cat. He speculated that the one from Texas might represent a different species, but cautiously—and correctly—labeled it as a mere variety. There are eleven races that differ slightly in size. Some are reddish-brown while others are pallid, but all of them—from the Atlantic to the Pacific and from lower Canada to central Mexico—are spotted, barred, and speckled with blackish markings that blend into brush or stony rubble.

PLATE I.

LYNX RUFUS, GULDENSTAED.

COMMON AMERICAN WILD CAT.

Lith. Printed & Colᵈ by J.T. Bowen Phila 1842.

Both bobcats and cougars are less partial to carrion than are most predators, and they must therefore make long forays in search of prey. The bobcat is an adaptable animal, comfortable in an upland forest, a riverbottom swamp, a brushy canyon, or the moraines above timberline. The cat at right balances easily on a dead limb while surveying a snowy ridge for hares. The one above is hunting pikas among fractured rocks on an alpine slope. Male bobcats are sexually active in all seasons; most females are receptive in March, but the time varies from one region to another. Gestation requires about two months. The kittens—most often three—are born in a rock den or hollow log, or sometimes beneath a windfall or in some other shelter. Weaned and taught to hunt during the summer, they can fend for themselves by midwinter.

No. 18
Drawn from Nature by J W Audubon
FELIS PARDALIS, LINN.
OCELOT, OR LEOPARD-CAT.
MALE

PLATE

86

Ocelot

Felis pardalis

Waiting for the sun to go down, an ocelot may groom its pair of kittens, bedded in a tree den or rock cavity, or it may stretch out on a limb to doze. Later it slips with fluid grace into the dense chaparral that hides rodents. Or perhaps it ripples furtively through treetops laden with perching birds, and in the tropics, with sleeping monkeys. Or it stalks the log-strewn bank of a stream, ready to plunge in and swim if need be to catch a reptile, fish, or amphibian. Mates sometimes hunt together, signaling each other with calls like the meows of a typical house cat that would be half their size and easy prey. Born at any time of year, ocelots are adapted to the hot climate prevailing from Mexico to Paraguay. They have never been abundant above the Mexican border, but in Texas and Arizona a few have eluded the fur smugglers, the protectors of lambs and poultry, and the bootleggers of exotic pets. An ocelot's pelt is richly patterned with black and sometimes brown, and no two specimens are alike. The markings are an asymmetric maze of spots, rosettes, rings, speckles, bars, and slashes against a sleek, soft grayish-gold.

During John Woodhouse Audubon's Texas expedition, while he was headquartered at San Antonio near an army encampment, General William Selby Harney presented to him a freshly killed ocelot from which he painted the plate for the *Quadrupeds*. "My delight was only equalled by my desire to paint a good figure," he said, for he ranked it as the most beautiful of North American felines. His father and John Bachman, similarly inspired by "the activity and grace of the Leopard-Cat," as the ocelot was sometimes called, wrote of "the beauty of its fur" and the way it "leaps with ease amid the branches . . . or runs with swiftness on the ground." In arid country where there are few trees, the slender ocelot moves with equally fluid agility through rocky canyons. Even the immature specimen in the top photograph, though it has not yet acquired the sleek fur of the adults in the other pictures, already displays the statuesque poise characteristic of the species.

Photographers Credits

NA National Audubon Collection/PR
PA Peter Arnold

OPOSSUM
24: (both) Leonard Lee Rue III, NA. 25: Leonard Lee Rue IV.

ARMADILLO
28-29: (top left) Leonard Lee Rue III; (bottom left) Bill Browning; (right) Ken Brate, NA.

LEAST SHREW
(and wandering shrew)
32-33: (left & top right) Stouffer Productions, Ltd., NA; (center right) Charles E. Mohr, NA; (bottom right) Cosmos Blanc, NA.

COMMON MOLE
36: Dr. Edward Denninger. 37: (both) Charles E. Mohr, NA.

COMMON COTTONTAIL
40-41: (top left) Steven Collins, NA; (bottom left & right) Tom Brakefield.

SWAMP RABBIT
44: (all) Kirtley-Perkins.
45: Thase Daniel.

VARYING HARE
48: (top) Charles Ott, NA; (bottom left) S. J. Krasemann, PA; (bottom right) William J. Jahoda, NA. 49: Charles Ott, NA.

WHITE-TAILED JACK RABBIT
(and black-tailed jack rabbit)
52: Bill Browning. 53: (top left) Kenneth Fink, NA; (bottom left) Leonard Lee Rue III, NA; (right) Leonard Lee Rue IV, NA.

PIKA
56: Charles Ott, NA.
57: (top) S. J. Krasemann, PA; (bottom) Bill Browning.

FLYING SQUIRREL
60: (top) Robert J. Erwin, NA; (bottom) Alvin E. Staffan, NA.
61: A. A. Francesconi, NA.

RED SQUIRREL
64: Larry West, NA.
65: (left) Ed Cesar, NA; (right) [illegible], NA.

FOX SQUIRREL
68: John H. Gerard, NA.
69: (left) Erwin A. Bauer; (right) Ed Cooper.

EASTERN GRAY SQUIRREL
72: Bill Browning.
73: Arthur Swoger.

WESTERN GRAY SQUIRREL
76: Bill Browning. 77: (top) Erwin A. Bauer; (bottom) C. M. Altimus, NA.

EASTERN CHIPMUNK
80: Larry West, NA.
81: (top) Alvin E. Staffan, NA; (bottom) Tom Brakefield.

THIRTEEN-LINED GROUND SQUIRREL
(and Richardson's and arctic ground squirrels) 84: (top) Bill Browning; (bottom left) Tom Brakefield; (bottom right) John R. MacGregor, PA. 85: Erwin A. Bauer.

BLACK-TAILED PRAIRIE DOG
88: Leonard Lee Rue III, NA.
89: Erwin A. Bauer.

WOODCHUCK
92: (top) Leonard Lee Rue III; (bottom) W. J. Schoonmaker, NA.
93: Tom Brakefield.

HOARY MARMOT
96: Kenneth Fink, NA.
97: (top) Erwin A. Bauer; (bottom) S. J. Krasemann, NA.

MOUNTAIN BEAVER
100-101: (all) Michael Wotton.

KANGAROO RAT
104: Anthony A. Mercieca, NA.
105: (top) Robert H. Wright, NA; (bottom) Alan Blank, NA.

PORCUPINE
108: (top) Leonard Lee Rue III, NA; (bottom) Edwin A. Park, NA.
109: Erwin A. Bauer.

BEAVER
112: Erwin A. Bauer. 113: (top) Leonard Lee Rue III, NA; (bottom left) Harry Engels, NA; (bottom right) A.A. Francesconi, NA.

MUSKRAT
116: (top) Charles Ott, NA; (center) A.A. Francesconi, NA; (bottom) Leonard Lee Rue III.
117: Leonard Lee Rue III, NA.

BROWN LEMMING
(and Labrador collared lemming)
120: D. Wilkinson, Information Canada Phototheque. 121: (top & bottom) Michael Wotton; (center) Roland C. Clement, NA.

WHITE-FOOTED MOUSE
(and deer mouse)
124: (top left) Arthur Swoger; (top right) Larry West, NA; (bottom) Leonard Lee Rue III.
125: Arthur Swoger.

MEADOW VOLE
128: Leonard Lee Rue III. 129: (both) John R. MacGregor, PA.

EASTERN WOOD RAT
(and bushy-tailed wood rat)
132: Stephen Collins, NA.
133: (top) James R. Tallon; (bottom) J. R. Simon, Bruce Coleman, Inc.

RINGTAIL
136-137: (left and bottom right) Robert J. Erwin, NA; (top right) Kenneth Fink, NA.

RACCOON
140-141: (left and top right) Tom Brakefield; (bottom right) Erwin A. Bauer.

BLACK BEAR
144: (top) Bill Browning; (bottom) Erwin A. Bauer. 145: Arthur Swoger.

GRIZZLY BEAR
(and Alaskan brown bear)
148: Bill Browning.
149: (top) S. J. Krasemann, NA; (bottom left & right) Tom Brakefield.

POLAR BEAR
152: (top) Erwin A. Bauer; (bottom) Ed Cooper. 153: Larry B. Jennings, NA.

RIVER OTTER
156: Ed Cooper. 157: (top) Erwin A. Bauer; (bottom) Leonard Lee Rue III, NA.

SEA OTTER
160: Erwin A. Bauer.
161: (top) Karl W. Kenyon, NA; (bottom) Marty Stouffer, NA.

PINE MARTEN
164: (top) National Film Board of Canada; (bottom) Marty Stouffer, NA. 165: Ed Cesar, NA.

FISHER
168-169: (both) Leonard Lee Rue III.

BLACK-FOOTED FERRET
172-173: (top & bottom right) B. J. Rose; (bottom left) South Dakota State University, Department of Wildlife and Fisheries Sciences.

LONG-TAILED WEASEL
(and short-tailed weasel)
176: Phil A. Dotson, NA.
177: (top) D. Mohrhardt, NA; (center) Charles Ott, NA; (bottom) Leonard Lee Rue III.

MINK
180: S. J. Krasemann, NA.
181: (top & bottom right) Leonard Lee Rue III, NA; (bottom left) H. A. Thornhill.

STRIPED SKUNK
(and spotted skunk)
184: (top & bottom) Leonard Lee Rue III, NA.
185: Erwin A. Bauer.

BADGER
188: (top) Leonard Lee Rue IV, NA; (bottom) Robert J. Erwin, NA.
189: Alvin E. Staffan, NA.

WOLVERINE
192: (top) Alan G. Nelson, NA; (bottom) Anthony Mercieca, NA.
193: S. J. Krasemann, PA.

GRAY FOX
196: Leonard Lee Rue III, NA.
197: Erwin A. Bauer.

RED FOX
200-201: Leonard Lee Rue III.

COYOTE
204: Harry Engels, NA. 205: (top) Joe Van Wormer, NA; (bottom) Erwin A. Bauer.

RED WOLF
208-209: (all) Marty Stouffer, NA.

GRAY WOLF
212: (top) Marty Stouffer, NA; (bottom) Dale P. Hansen, NA.
213: Erwin A. Bauer.

BISON
216: (top right & left) Bill Browning; (bottom) Erwin A. Bauer. 217: Bill Browning.

PRONGHORN
220-221: Erwin A. Bauer

MOUNTAIN GOAT
224: Phil Farnes, NA.
225: (top) John M. Burnley; (bottom left) Leonard Lee Rue III; (bottom right) Bill Browning.

BIGHORN SHEEP
228-229: (left) Erwin A. Bauer; (top right) Bill Browning; (bottom right) Tom Brakefield.

WHITE-TAILED DEER
232: (top & bottom) Bill Browning; (center) Erwin A. Bauer. 233: Tom Brakefield.

MULE DEER
236: Erwin A. Bauer.
237: (top left) Leonard Lee Rue IV, NA; (top right) Bill Browning, (bottom) Erwin A. Bauer.

COLUMBIAN BLACK-TAILED DEER
240: Helen Cruickshank, NA. 241: (top) Kenneth Fink, NA; (bottom) Leonard Lee Rue IV, NA.

ELK
244-245: (top left) Ed Cooper; (bottom left & right) Erwin A. Bauer.

MOOSE
248: Charles Ott, NA.
249: Erwin A. Bauer.

CARIBOU
252-253: Charles Ott, NA.

COLLARED PECCARY
256: (top) Leonard Lee Rue III, (bottom) Kenneth Fink, NA. 257: Bill Browning

COUGAR
260: Tom Brakefield. 261: (top) John S. Crawford, NA. (bottom) Erwin A. Bauer.

BOBCAT
264-265: (left) Marty Stouffer, NA; (right) Larry West, NA.

OCELOT
268: Kenneth Fink, NA. 269: (top) Robert C. Hermes, NA; (bottom) Marty Stouffer, NA.

Index

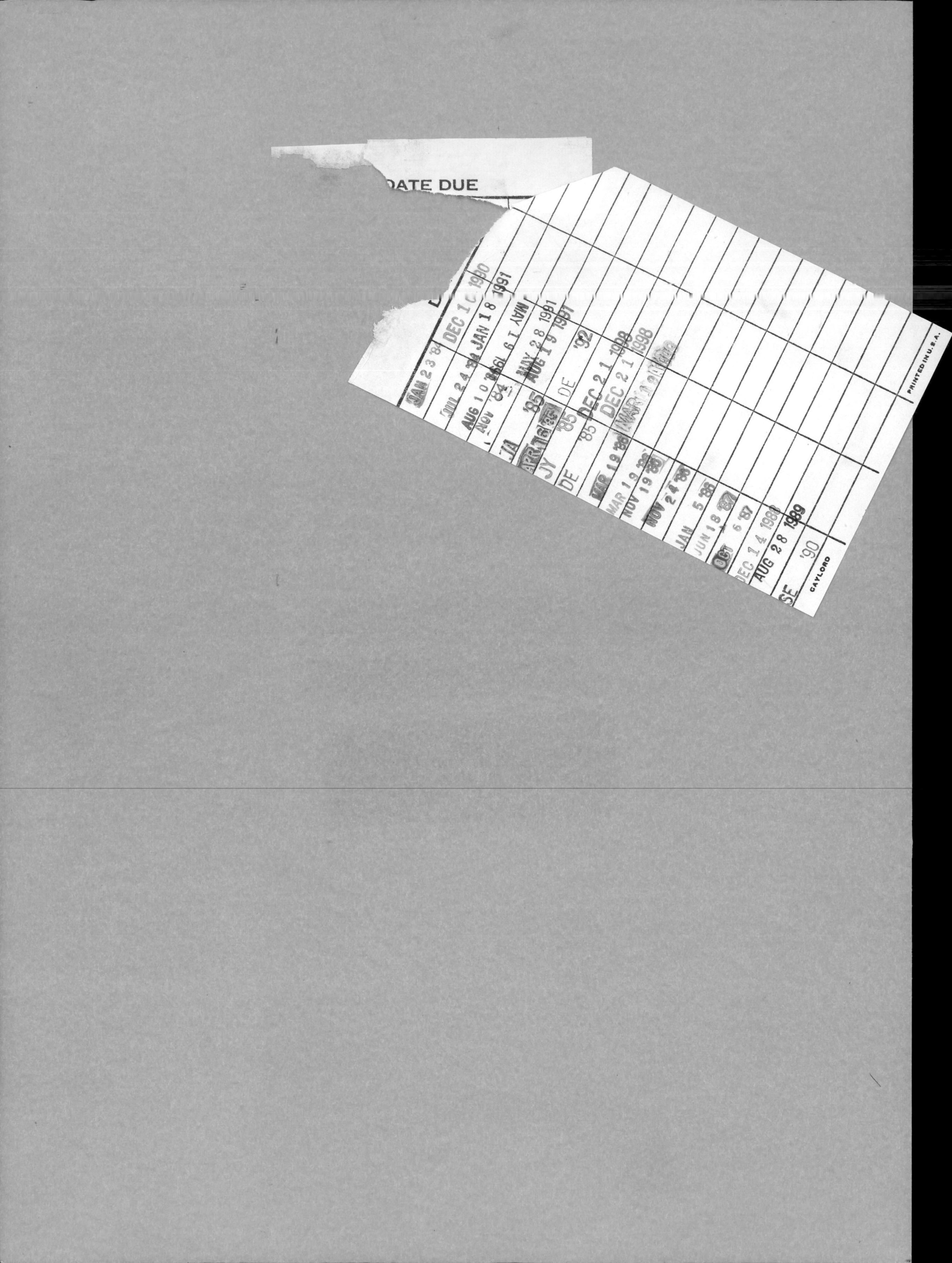